IMPROVE YOUR Math

IMPROVE YOUR Math

Problem Solving Made Simple for Middle School Math

Francine D. Galko

NEW YORK

Library of Congress Cataloging-in-Publication Data
Galko, Francine.
Improve your math : problem solving made easy for the middle school years / Francine Galko.—1st ed.
p. cm.
ISBN 1-57685-406-X
1. Mathematics—Study and teaching (Middle school) I. Title

QA11.2 .G35 2002
513'.14—dc21

2001038994

Printed in the United States of America

9 8 7 6 5 4 3 2 1

First Edition

ISBN 1-57685-406-X

For more information or to place an order, contact
LearningExpress at:
900 Broadway
Suite 604
New York, NY 10003

Or visit us at:
www.learnatest.com

Contents

Introduction

Middle school is an exciting time of your life. It's full of changes and challenges. By eighth grade, you are expected to be proficient in a number of areas—you'll be required to take tests that measure your reading, writing, and math skills. So, take the time now to brush up your math skills before you hit high school.

▶ HOW THIS BOOK CAN HELP YOU

If you're having trouble in math, this book can help you get on the right track. Even if you feel pretty confident about your math skills, you can use this book to help you review what you've already learned and practice with the same kinds of questions you'll see on standardized and classroom math tests, so you are sure to be prepared for the coming challenges of high school.

Many study guides tell you how to improve your math—this book doesn't just *tell* you how to solve math problems, it *shows* you how. You'll find page after page of strategies that work, and you are never left stranded wondering how to get the right answer to a problem. You are shown all the steps to take so that you can successfully solve every single problem, and see the strategies at work. This book is like your own personal math tutor!

With help from this book, you will be prepared for the higher-level mathematics you'll encounter in high school. So, now's the time to improve your math skills because you will use them—both now and in the future—in school, at work, and in your personal life. As you work through this book, you will become more confident in math and you can pass standardized tests and classroom math tests more easily.

HOW THIS BOOK IS ORGANIZED

This book is organized into eight sections, or major areas of math, such as fractions, decimals, algebra, and geometry. Each section is divided into short lessons that explain the main concepts you need to know. Each lesson tells you what kinds of key words and problems to look for when you are solving math problems, and gives you step-by-step strategies to solve each kind of problem you will encounter.

Sometimes math books assume you can follow explanations of difficult concepts even when they move quickly or leave out steps. That's *not* the case with this book. In each lesson, you'll find step-by-step strategies for tackling the different kinds of math problems you are likely to find on standardized tests and classroom math tests. Then, you're given a chance to apply what you've learned with hundreds of practice problems. Answers to the practice problems are provided at the end of each section, so you can check your progress as you go along.

As you work through this book, you'll notice that the lessons are sprinkled with all kinds of helpful tips and icons. Look for these icons and the tips they provide.

Shortcut When you see this icon, you'll know that a quicker, easier way to solve a math problem follows. You'll find this icon both in the lessons and in the answer key, so you can practice the shortcuts and see how they work in real problems.

Test Taking Tip This icon tells you about tips especially useful for taking math tests. They cover additional strategies or well-organized information you can use as a reference, long after you've worked through all the lessons in the book.

Think About It This icon is used to signal another way of looking at a problem or concept. You'll get extra information and more in-depth discussions when you see this icon.

Real World Problems Each section ends with a set of Real World Problems. These aren't the same old practice problems you'll find in other math books. These problems apply the skills you've learned in each section to situations you encounter every day. As you work through these problems, you'll see that the skills you're learning aren't important only for math tests. They are also important skills to know for questions that come up every day.

A WORD ABOUT WORD PROBLEMS

If you're looking specifically for help with word problems, you've come to the right place! Every section of this book includes word problems and step-by-step strategies for solving them. Not only do most of the lessons end with word problems, but every single section ends with a set of word problems that you'll encounter in the real world.

In addition to including lots of word problems for you to practice on, these Real World Problem sections include test-taking tips for tackling all kinds of word problems. That's not all—the answer key will walk you through solving every word problem. You'll feel like you have your own personal tutor.

HOW TO USE THIS BOOK

You should get started by taking the Pretest. You'll test your math skills and see where you might need to focus your study.

Then, begin the lessons, study the example problems, and try your hand at the practice problems. Check your answers as you go along, so if you miss a question, you can study a little more before moving on to the next lesson.

After you've completed all 21 lessons in the book, try your hand at the Posttest to see how much you've learned. You'll also be able see any areas where you may need a little more practice. You can go back to the section that covers that skill for some more review and practice.

Good luck, and let's get started!

Getting Started

▶ PRETEST

Before you start Lesson 1, you might want to take this short Pretest. The Pretest is 25 multiple-choice questions that cover the basic concepts and skills covered in this book. Of course, every skill, shortcut, concept, and type of problem in the book cannot be covered in only 25 questions; however, the Pretest can give you a general idea of which skills you know well and which skills you might need to work on.

You can use the answer sheet on the next page to record your answers to the Pretest. Or, you can simply circle your answers as you work through the Pretest. If this book doesn't belong to you, write the numbers 1-25 on a sheet of your own paper and record your answers there.

Take your time to work through the Pretest. Don't feel rushed. It may take a few questions to get your mind thinking about math. When you have completed the Pretest, check your answers against the answer key at the end of the test.

PRETEST ANSWER SHEET

1. ⓐ ⓑ ⓒ ⓓ
2. ⓐ ⓑ ⓒ ⓓ
3. ⓐ ⓑ ⓒ ⓓ
4. ⓐ ⓑ ⓒ ⓓ
5. ⓐ ⓑ ⓒ ⓓ
6. ⓐ ⓑ ⓒ ⓓ
7. ⓐ ⓑ ⓒ ⓓ
8. ⓐ ⓑ ⓒ ⓓ
9. ⓐ ⓑ ⓒ ⓓ
10. ⓐ ⓑ ⓒ ⓓ
11. ⓐ ⓑ ⓒ ⓓ
12. ⓐ ⓑ ⓒ ⓓ
13. ⓐ ⓑ ⓒ ⓓ
14. ⓐ ⓑ ⓒ ⓓ
15. ⓐ ⓑ ⓒ ⓓ
16. ⓐ ⓑ ⓒ ⓓ
17. ⓐ ⓑ ⓒ ⓓ
18. ⓐ ⓑ ⓒ ⓓ
19. ⓐ ⓑ ⓒ ⓓ
20. ⓐ ⓑ ⓒ ⓓ
21. ⓐ ⓑ ⓒ ⓓ
22. ⓐ ⓑ ⓒ ⓓ
23. ⓐ ⓑ ⓒ ⓓ
24. ⓐ ⓑ ⓒ ⓓ
25. ⓐ ⓑ ⓒ ⓓ

1. Anne has two containers for water: A rectangular plastic pitcher with a base of 16 square inches, and a cylindrical vase with a radius of 2 inches and a height of 11 inches. If the rectangular pitcher is filled with water 9 inches from the bottom, and Anne pours the water into the vase without spilling, which of the following will be true?

a. The cylinder will overflow.
b. The cylinder will be exactly full.
c. The cylinder will be filled to an approximate level of 10 inches.
d. The cylinder will be filled to an approximate level of 8 inches.

2. What fraction of the figure is shaded in?

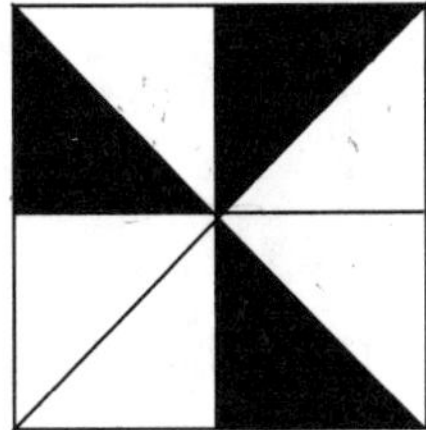

a. $\frac{1}{2}$
b. $\frac{1}{4}$
c. $\frac{2}{3}$
d. $\frac{3}{8}$

3. Write $\frac{12}{4}$ as a whole number.

a. 8
b. 4
c. 3
d. 2

Use the chart below to answer Questions 4 and 5.

LAURIE'S JOGGING LOG

DAY	MILES JOGGED
Sunday	4
Monday	3
Tuesday	$3\frac{1}{3}$
Wednesday	5
Thursday	$2\frac{1}{3}$
Friday	$2\frac{2}{3}$
Saturday	3

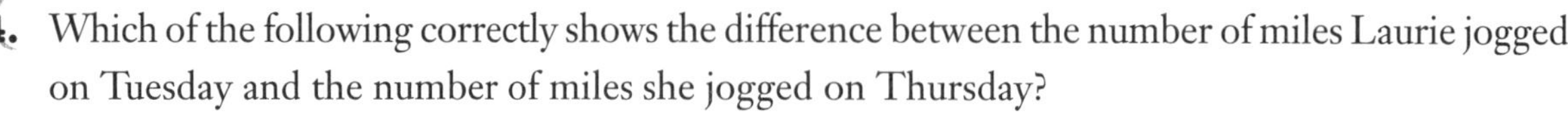

4. Which of the following correctly shows the difference between the number of miles Laurie jogged on Tuesday and the number of miles she jogged on Thursday?

a. $3\frac{1}{3} + 2\frac{1}{3}$

b. $3 - 2\frac{1}{3}$

c. $3\frac{1}{3} - 2\frac{2}{3}$

d. $3\frac{1}{3} - 2\frac{1}{3}$

5. What was Laurie's average daily jog for the week shown?

a. $2\frac{2}{3}$ miles per day

b. 3 miles per day

c. $3\frac{1}{3}$ miles per day

d. $3\frac{2}{3}$ miles per day

6. Rayeel jogs at an average rate of $5\frac{1}{3}$ miles per hour. If he just jogged 16 miles, how long has he jogged so far?

a. 3 hours

b. $3\frac{1}{3}$ hours

c. $3\frac{2}{3}$ hours

d. 4 hours

7. Faleena was given a 110-question homework assignment. On Friday night, she did $\frac{1}{5}$ of the assignment. On Saturday, she finished $\frac{1}{4}$ of the remaining questions. On Sunday, she did another $\frac{1}{3}$ of the remaining questions. How many questions does Faleena have left?

a. 23

b. 32

c. 44

d. 51

8. Adam collected the following amounts of change from under his living room furniture: 0.24, 0.75, 0.89, and 1.27. How much money did he find altogether?

a. $315.00

b. $31.50

c. $3.15

d. $0.315

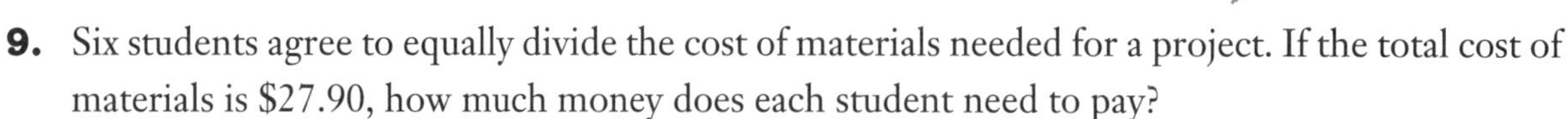

9. Six students agree to equally divide the cost of materials needed for a project. If the total cost of materials is $27.90, how much money does each student need to pay?

a. $4.50

b. $4.65

c. $6.75

d. $6.80

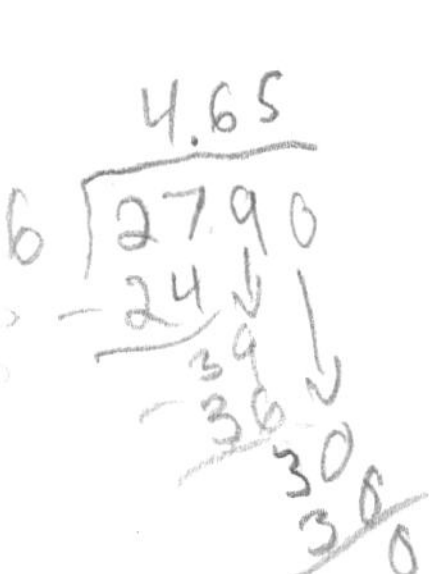

10. Which of the following percents is equivalent to the fraction $\frac{42}{50}$?

a. 42%

b. $\frac{42}{50}$%

c. 84%

d. 90%

11. Which of the following is 17% of 3,400?

a. 200

b. 340

c. 578

d. 620

The pie chart below shows Charlie's monthly expenses.
Use this information to answer Question 12.

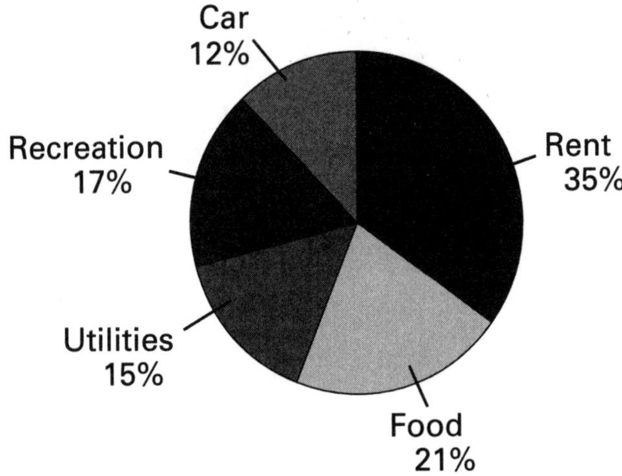

12. If Charlie's total expenses are $1,500 per month, how much does he spend on food over the course of *two* months?

a. $270

b. $315

c. $420

d. $630

13. A scanner that usually sells for $448 is on sale for 15% off. Which of the following represents the sale price of the scanner?

a. $448 – 15

b. 0.15 × $448

c. $448 – (0.15 × $448)

d. $448 × 15

14. Which of the following represents $3x + 15 = 32$?

a. 15 less than 3 times a number is 32
b. 32 times 3 is equal to 15 more than a number
c. 15 more than 3 times a number is 32
d. 3 more than 15 times a number is 32

15. When 42 is subtracted from a number, the result is 56. What is the number?

a. 14
b. 28
c. 54
d. 98

16. Look at the figure below. What is the value of y?

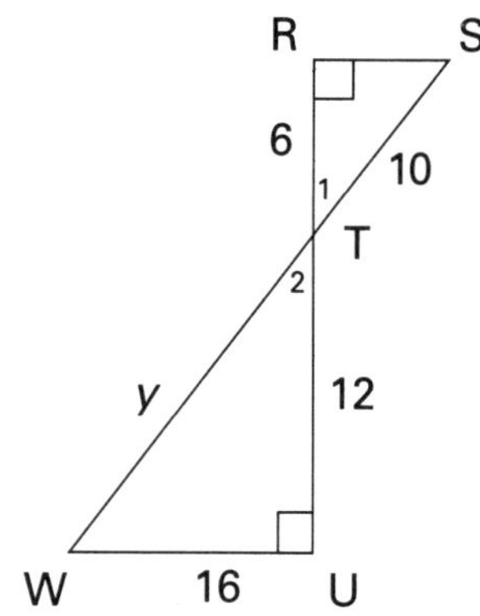

a. 15
b. 20
c. 30
d. 40

17. What is the measure of angle 1 in the figure below?

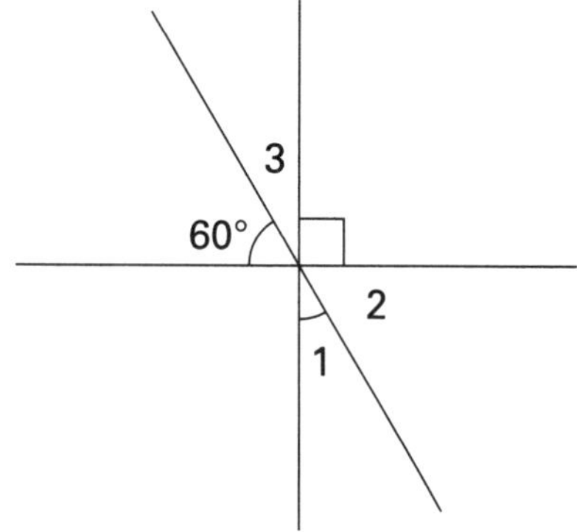

a. 15°
b. 30°
c. 45°
d. 90°

18. $5 \times 2 + 16 \div 4 =$

a. 6.5
b. 14
c. 20
d. 24

19. What is the length of one side of a square that has an area of 324 square inches?

a. 18 inches
b. 18^2 inches
c. 19 inches
d. $\sqrt{19}$ inches

20. A sack holds 3 purple buttons, 2 orange buttons, and 5 green buttons. What is the probability of drawing one purple button out of the sack and then—without replacing the first button—drawing a second purple button out of the sack?

a. $\frac{1}{15}$
b. 15%
c. $\frac{47}{90}$
d. $\frac{6.7}{100}$

21. What is the length of $\overline{DC}$ in the figure below?

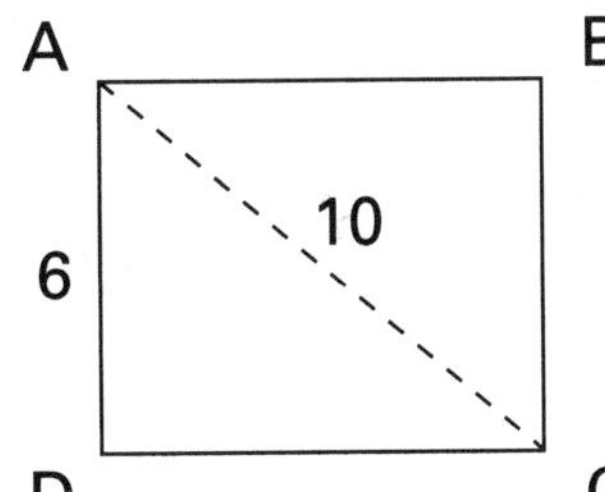

a. 6
b. 8
c. 10
d. 12

22. A bag contains 4 red tokens, 5 yellow tokens, and 3 blue tokens. What is the probability that the first token drawn at random from the bag will be red?

a. $\frac{1}{3}$
b. $\frac{1}{4}$
c. $\frac{1}{5}$
d. $\frac{1}{12}$

23. What is the perimeter of the triangle shown below?

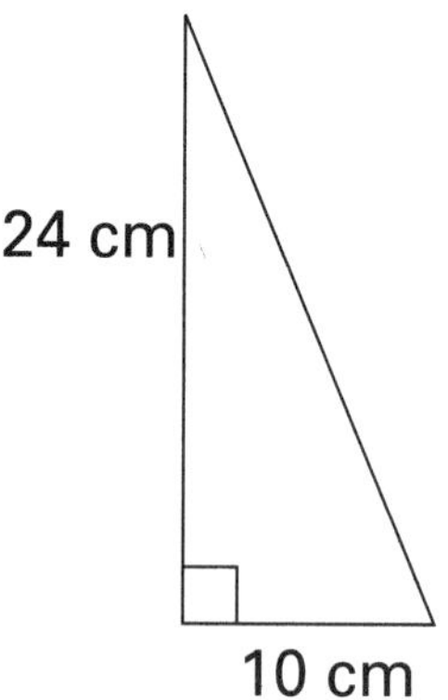

a. 26 cm
b. 37 cm
c. 57 cm
d. 60 cm

24. How many miles would you drive if you started in Los Angeles, drove to Chicago, and then drove to Boston?

U.S. CITY MILEAGE TABLE

	Atlanta	Boston	Chicago	Denver	Los Angeles
Atlanta		1037	674	1398	2182
Boston	1037		994	1949	2979
Chicago	674	994		996	2054
Denver	1398	1949	996		1059
Los Angeles	2182	2979	2054	1059	

a. 994
b. 2,054
c. 2,979
d. 3,048

25. Evaluate $m = n\,(10 - 3) + (15 - n)$, when $n = 3$.

a. 27
b. 30
c. 33
d. 36

PRETEST ANSWERS AND EXPLANATIONS

1. a. Begin by calculating the volume of the water. Use the formula for the volume of a rectangular solid to calculate the volume of the water: $V = Ah = lwh$. The area of the base is 16 inches^2 and the height of the water is 9 inches, so to get the volume of the water in the pitcher you plug these known measures into the formula and solve:

$V = Ah = lwh$
$V = (16 \text{ in}^2)(9 \text{ in})$
$V = 144 \text{ in}^3$

So, the volume of the water is 144 in^3.
Next, calculate the volume of the vase. The formula for the volume of a cylinder is $V = Ah$ or $V = \pi r^2h$. So, you plug in the measures and solve:

$V = \pi r^2h$
$V = \pi(2)^2(11)$
$V = \pi(4)(11)$
$V = \pi(44)$
$V = 138.16$

So the volume of the vase is about 138 cm^3, and this volume is smaller than the volume of water in the pitcher, so the vase will overflow if Anne tries to fill it with all the water in the pitcher. These concepts are covered in Lesson 19.

2. d. Because 3 out of 8 total parts are shaded, you know that $\frac{3}{8}$, or three-eighths, of the whole figure is shaded. This concept is covered in Lesson 3.

3. c. To change the improper fraction $\frac{12}{4}$ to a whole number, just divide 12 by 4: $12 \div 4 = 3$. So, your answer is 3. This concept is covered in Lesson 6.

4. d. Look at the chart and find the number of miles jogged on Tuesday and Thursday. Laurie jogged $3\frac{1}{3}$ miles on Tuesday and $2\frac{1}{3}$ miles on Thursday, so you subtract: $3\frac{1}{3} - 2\frac{1}{3}$. The answer is $3\frac{1}{3} - 2\frac{1}{3}$. These concepts are covered in Lessons 4 and 15.

5. c. The formula for finding an average (or mean) is as follows:

$$\text{Average} = \frac{\text{the sum of the numbers}}{\text{the number of numbers}}$$

So you need to begin by finding the total number of miles Laurie jogged for the week given: $4 + 3 + 3\frac{1}{3} + 5 + 2\frac{1}{3} + 2\frac{2}{3} + 3$. When adding mixed numbers, first add the whole numbers, then add the fractions. Next, combine the whole number sum with the fraction sum. If you add the whole numbers, you get $4 + 3 + 3 + 5 + 2 + 2 + 3 = 22$. Adding the fractions gives you $\frac{1}{3} + \frac{1}{3} + \frac{2}{3} = \frac{4}{3}$. The improper $\frac{4}{3}$ converts to the mixed number $1\frac{1}{3}$. So you add 22 and $1\frac{1}{3}$ to get $23\frac{1}{3}$ miles. Now you are ready to plug the numbers into the average formula:

$$\text{Average} = \frac{23\frac{1}{3}}{7}$$

The fraction bar is a division sign, so you can simplify this problem by writing it as the following division problem: $23\frac{1}{3} \div 7$. You can further simplify your division by converting the mixed number to an improper fraction. Multiply 23 × 3, add 1, and then put this over 3:

$$23 \xrightarrow{\text{times}} \frac{1}{3} \text{ plus}$$

$$= \frac{70}{3}$$

Now you are ready to divide: $\frac{70}{3} \div 7$. Write the whole number 7 as a fraction ($\frac{7}{1}$), invert it ($\frac{1}{7}$) and multiply: $\frac{70}{3} \times \frac{1}{7}$. Using canceling, you can simplify the multiplication:

$$\frac{\cancel{70}^{10}}{3} \times \frac{1}{\cancel{7}_{1}} = \frac{10}{3}$$

The improper fraction $\frac{10}{3}$ is equal to the mixed number $3\frac{1}{3}$. So your final answer is $3\frac{1}{3}$ miles per day. These concepts are covered in Lessons 3–6 and 11.

6. a. Begin by setting up a proportion:

$$\frac{5\frac{1}{3}\text{ miles jogged}}{1\text{ hour}} = \frac{16\text{ miles jogged}}{?\text{ hours}} = \frac{5\frac{1}{3}}{1} = \frac{16}{x}$$

Then, cross multiply and solve for x. You can simplify the multiplication by first converting the mixed number $5\frac{1}{3}$ to the improper fraction $\frac{16}{3}$.

$$\frac{16}{3}x = 16$$

$$\frac{3}{16} \times \frac{16}{3}x = 16 \times \frac{3}{16}$$

$$x = 3$$

So Rayeel has been jogging for 3 hours. These concepts are covered in Lessons 3–6, and 12.

7. c. When you are asked to take a fraction *of* another number, you are being asked to *multiply* the number by that fraction. Here's what Faleena does each day:

Day	Number of Questions
Start	110
Friday	minus $\frac{1}{5} \times 110$, or −22 = 88
Saturday	minus $\frac{1}{4} \times 88$, or −22 = 66
Sunday	minus $\frac{1}{3} \times 66$, or −22 = 44

So Faleena still has 44 questions left on her assignment. These concepts are covered in Lesson 5.

8. c. To solve this problem correctly, you must be sure that the decimal points are lined up correctly. Now you are ready to add:

$$\begin{array}{r} 0.24 \\ 0.75 \\ 0.89 \\ +1.27 \\ \hline 3.15 \end{array}$$

So, Adam found \$3.15. This concept is covered in Lesson 7.

9. b. Divide the total amount (\$27.90) by six students. To do this, you need to first set up the division problem. Begin by writing the decimal point in the answer portion of the division problem. It should go just above the decimal point in the problem. Now you are ready to divide:

$$6\overline{)27.90}$$

Your final answer is \$4.65. These concepts are covered in Lesson 8.

10. c. When written as fractions, percents have a denominator of 100. You can convert $\frac{42}{50}$ to a fraction with a denominator of 100 by multiplying by $\frac{2}{2}$: $\frac{2}{2} \times \frac{42}{50} = \frac{84}{100}$. Then, write the numerator alone and add a percent sign to get 84%. These concepts are covered in Lesson 9.

11. c. Begin by figuring out what you know from the problem and what you're looking for.

- You have the percent: 17%.
- You have the whole: 3,400.
- You are looking for the part.

Then, use the following equation to solve the problem: Whole × Percent = Part. Plug in the parts of the equation that you know: Part = 3,400 × 0.17 = 578. The answer is 578. These concepts are covered in Lesson 10.

12. d. According to the graph, 21% of total monthly expenses go toward food, so you need to find 21% of \$1,500. Twenty-one percent is equal to 0.21 and *of* means to multiply: 0.21 × 1,500= 315. So Charlie spends \$315 on food each month. But the problem asks how much he spends on food in *two* months, so you need to multiply 315 by 2 to get your final answer: \$630. These concepts are covered in Lesson 14.

13. c. To find the sale price of the scanner, you need to subtract the discount (15% of the regular price) from the regular price. So the correct answer is \$448 – (0.15 × \$448). These concepts are covered in Lesson 10.

14. c. Break the math symbols down as follows:

$$\underbrace{3x}_{\text{3 times a number}} + \underbrace{15}_{\text{plus 15}} \underbrace{= 32}_{\text{equals 32}}$$

So the answer is 15 more than 3 times a number is 32. These concepts are covered in Lessons 16 and 17.

15. d. Begin by letting x = the number you are looking for. Then, set up an algebraic equation and solve for x. From the problem, you know that subtracting 42 from x gives you 56, so

$$x - 42 = 56$$

Then, add 42 to each side of the equation to solve for x.

$$\begin{array}{rcl} x - 42 & = & 56 \\ +42 & & +42 \\ \hline x & = & 98 \end{array}$$

So the final answer is 98. These concepts are covered in Lesson 17.

16. b. You know that the triangles ΔRST and ΔUWT are similar because two of their angles are congruent. Both triangles have right angles and angles 1 and 2 are vertical angles (which are congruent). The sides of similar triangles are proportional. So you can set up a proportion to solve for the missing length.

$$\frac{6}{12} = \frac{10}{y}$$

Cross multiply to solve for y.

$$6y = 10 \times 12$$
$$\frac{6y}{6} = \frac{120}{6}$$
$$y = 20$$

So, y = 20. These concepts are covered in Lesson 20.

17. b. From the diagram, you know that the sum of angles 1 and 2 is 90°. You also know that angle 2 is 60° because it is vertical to (or opposite of) another 60° angle. To calculate the measure of angle 1, you simply subtract: 90 – 60 = 30. So the answer is 30°. These concepts are covered in Lesson 18.

18. b. To solve this problem, you simply follow the order of operations. First, multiply and divide numbers in order from left to right. You multiply 5 by 2, then divide 16 by 4:

$$5 \times 2 + 16 \div 4 = 10 + 4$$

Then, add and subtract numbers in order from left to right:

$$10 + 4 = 14$$

So, the final answer is 14. These concepts are covered in Lesson 1.

19. a. The problem is asking you to calculate the square root of 324 because the area of a square is *side* × *side*, or $side^2$. So, you ask yourself what number multiplied by itself equals 324: 18 × 18 is 324. Thus, the square root of 324 is 18, and the length of one side of a square with an area of 324 square inches is 18 inches. These concepts are covered in Lesson 2.

20. a. First, determine the probability that each event will occur. Notice that the first event—drawing a purple button out of the sack—affects the probability of the second event because it changes both the number of purple buttons still in the sack, as well as the total number of buttons in the sack. The probability of drawing the first purple button is $\frac{3}{10}$, and the probability of drawing a second purple button is $\frac{2}{9}$. Then, you multiply the two probabilities together to find the probability that the two events will occur:

$$\frac{3}{10} \times \frac{2}{9} = \frac{6}{90} = \frac{1}{15}$$

Therefore, the probability of drawing two purple buttons consecutively is $\frac{1}{15}$. These concepts are covered in Lesson 13.

21. b. Notice that ACD forms a right triangle. So you can use the Pythagorean Theorem ($a^2 + b^2 = c^2$) to solve for the missing length. First, match up the variables a, b, and c with the parts of the right triangle. For example, you might let a = the length of $\overline{AD}$ and b = the length of $\overline{DC}$. The variable c must equal the length of $\overline{AC}$ because it is the hypotenuse. Next, plug the known lengths into the Pythagorean Theorem and solve.

$$a^2 + b^2 = c^2$$
$$6^2 + b^2 = 10^2$$
$$36 + b^2 = 100$$
$$\begin{array}{r} 36 + b^2 = 100 \\ -36 \qquad -36 \\ \hline b^2 = 64 \end{array}$$
$$\sqrt{b^2} = \sqrt{64}$$
$$b = 8$$

So, the length of the missing leg is 8. These concepts are covered in Lesson 21.

22. a. There are four red tokens (favorable outcomes) and twelve total tokens (possible outcomes). Plug these numbers into the equation for probability:

$$\text{P (event)} = \frac{\text{number of favorable outcomes}}{\text{number of total outcomes}} = \frac{4}{12}$$

Therefore, the probability of drawing a red token is $\frac{4}{12}$, which reduces to your final answer, $\frac{1}{3}$. These concepts are covered in Lesson 13.

23. d. First, determine the missing length of the hypotenuse of the right triangle. Since the triangle is a right triangle, you can use the Pythagorean Theorem to solve for the missing length. Plug the variables into the equation and solve for c:

$$a^2 + b^2 = c^2$$
$$24^2 + 10^2 = c^2$$
$$576 + 100 = c^2$$
$$676 = c^2$$
$$\sqrt{676} = \sqrt{c^2}$$
$$26 = c$$

So the length of the hypotenuse is 26 cm. Now you are ready to calculate the perimeter of the triangle: 24 + 10 + 26 = 60. So your final answer is 60 cm. These concepts are covered in Lessons 19 and 21.

24. d. Los Angeles to Chicago is 2,054 miles. Chicago to Boston is 994 miles. Add the two: 2,054 + 994 = 3,048. So you would drive 3,048 miles. These concepts are covered in Lesson 15.

25. c. To evaluate an algebraic expression means to find its value by plugging in the known values of its variables. So begin by replacing the variable n with the number 3:

$m = n\ (10 - 3) + (15 - n)$

$= 3\ (10 - 3) + (15 - 3)$

Follow the order of operations and solve for m.

$= 3\ (10 - 3) + (15 - 3)$

$= 3\ (7) + (12)$

$= 21 + 12$

$= 33$

Thus, the answer is 33. These concepts are covered in Lessons 1 and 16.

SECTION

1

Whole Numbers

YOU HAVE PROBABLY covered these basic number operations and number concepts in previous math studies, but the concepts in this section are important facts that need to be practiced and reinforced. For all students, a clear understanding of these basic skills is a necessary foundation on which to build math proficiency. In this section, you'll learn how to use what you already know and apply that knowledge to solve math problems of any sort.

Order of Operations

LESSON SUMMARY

Many math problems require you to do more than one calculation. This lesson will show you how to determine which calculations to do and the order to do them in. This lesson will also help you review adding, subtracting, multiplying, and dividing whole numbers.

W*hole numbers* are made up of ten digits: 0, 1, 2, 3, 4, 5, 6, 7, 8, and 9. In this lesson, you will work only with whole numbers. In later lessons, you will learn specific ways to deal with numbers that come in between whole numbers. These numbers include 6.5, $\frac{1}{2}$, 34.6, $\frac{2}{3}$, and so on.

SOLVING PROBLEMS WITH MULTIPLE STEPS

You are familiar with the four basic *operations*, or ways of calculating: adding, subtracting, multiplying, and dividing. Sometimes a problem will ask you to do more than one operation. For example, if you are asked to solve this problem, what should you do?

$8 \times 3 + 20 \div 4 =$

You could do the operations in order from left to right. That is, you could multiply ($8 \times 3 = 24$), add ($24 + 20 = 44$), then divide ($44 \div 4 = 11$) to get 11. But you would not get the correct answer. The correct answer is 29. It looks tricky, but it's not if you know the *order of operations*. The order of operations involves three simple steps. When you follow these steps, you will get the correct answer.

THE ORDER OF OPERATIONS

Step 1: Do all the operations in parentheses.
Step 2: Multiply and divide numbers in order from left to right.
Step 3: Add and subtract numbers in order from left to right.

Example: $2 + 5 - (9 \div 3) \times 2 =$
To solve this problem, you should follow the steps in the table above.

Step 1: Do the operations in parentheses first.
$2 + 5 - (9 \div 3) \times 2 =$
$2 + 5 - (3) \times 2 =$
Step 2: Multiply.
$2 + 5 - 6 =$
Step 3: Add and subtract numbers in order, from left to right.
$7 - 6 = 1$

If you have a series of numbers to add or multiply, the order will not affect your final answer. You can group the numbers in a way that makes the addition or multiplication easier.

Examples:

$3 + 6 = 6 + 3$
$9 \times 2 = 2 \times 9$
$(2 + 3) + 5 = 2 + (3 + 5)$
$4 \times (6 \times 8) = (4 \times 6) \times 8$

So, if you were asked to solve the following problem

$27 + 5 + 3 + 15 =$

you might group 27 + 3 and 5 + 15 to make the math easier and faster. Do you see how grouping can make a problem easier? How could you group the numbers in the following problem to make it easier?

$12 \times 7 \times 5 =$

continued from previous page

If you know that 12 × 5 is 60, you could do this calculation first. Then calculate 60 × 7 (420). Notice that if you did 12 × 7 first, then you would end with 84 × 5, which isn't as quick to calculate at 60 × 7. Regrouping the numbers can speed up your calculations.

Keep in mind that *all* of the operations in the series must be *either* addition *or* multiplication for this shortcut to work. Also, remember that the order of the numbers in subtraction and division *is* important. You cannot change the order of subtraction and division numbers and still get the correct answer.

PRACTICE

Solve the following problems using the order of operations. You can check your answers at the end of the section.

1. (8 + 2) – 3 × 2 =
2. 9 × 5 + 3 ÷ 1 =
3. (3 + 4) × (2 + 6) =
4. (8 + 12) + (6 ÷ 2) =
5. 9 ÷ 3 + 7 =
6. 9 × 7 + 8 ÷ 4 =
7. 6 × (5 + 2) – 1 =
8. (10 × 4) + 12 – 6 =
9. (9 + 3) × (18 ÷ 3) =
10. (18 + 6) ÷ (18 – 12) =
11. 5 × 7 + 16 ÷ 4 =
12. 12 + 8 – (20 × 2) ÷ 10 =
13. 3 × 9 – 15 ÷ 5 =
14. 14 – 1 – 4 ÷ 2 =

15. $12 + 4 \div 4 \times 4 + 7 =$

16. $(13 + 2) \div 3 + 2 =$

17. $11 + 5 + 4 \times 3 + 7 =$

18. $8 \times 6 + 10 \div 2 =$

19. $4 \times 10 - 7 + 17 + 7 \times 2 =$

20. $8 \times 4 + (21 \div 3) - 7 + 9 - 1 =$

▶ CHOOSING AN OPERATION

Often a problem will tell you exactly which operation you should do. However, sometimes you will have to translate the words in a word problem into the operations. Look for these clues when you have to choose the operations.

You add (+) when you are asked to

- ▶ find a sum
- ▶ find a total
- ▶ combine amounts

Key words to look for:

- ▶ sum
- ▶ total
- ▶ altogether

You subtract (–) when you are asked to

- ▶ find a difference
- ▶ take away an amount
- ▶ compare quantities

Key words to look for:

- ▶ difference
- ▶ take away
- ▶ how many more than
- ▶ how much less than
- ▶ how many fewer than
- ▶ how much is left over

You multiply (×, ·) when you are asked to

- ▶ find a product
- ▶ add the same number over and over

Key words to look for:

- product
- times

You divide (÷) when you are asked to

- find a quotient
- split an amount into equal parts

Key words to look for:

- quotient
- per

Example: Add the product of 6 and 3 to the sum of 10 and 4.
To solve this problem, begin by translating the words into math symbols. You know from the lists on the previous page and above on this page that the word *product* means to multiply. So you will need to multiply 6 and 3. You also know that *sum* means to add. Thus, you could write the problem like this:

$6 \times 3 + 10 + 4 =$

Now follow the order of operations to solve the problem you have written:

Step 1: There are no parentheses. Skip to Step 2.
Step 2: Multiply.
$6 \times 3 + 10 + 4 =$
$18 + 10 + 4 =$
Step 3: Add in order from left to right.
$28 + 4 = 32$

Example: Elsa and Thuy went to a movie at the cinema. They shared a large popcorn. Each girl paid for her own drink. The movie cost $6.25. The popcorn cost $4.50. Each drink cost $2. How much did each girl pay?

Begin by translating the words into math symbols. The cost of the popcorn should be divided between the two girls. So, each girl paid

$6.25 + (4.50 \div 2) + 2 =$

Now solve the problem following the order of operations.

Step 1: Do operations in parentheses first.
$6.25 + (2.25) + 2 =$
Step 2: There is no multiplication or division. Skip to Step 3.
Step 3: Add.
$8.5 + 2 = 10.50$

Each girl paid $10.50 for the movie and food.

What if the question had asked: How much did the two girls pay altogether? How would you write the problem in math symbols to answer this question? There is more than one way to write it. Here are some ways you might recognize:

$$2 \times (6.25 + 2) + 4.50 =$$
$$2 \times 6.25 + 4.50 + 2 \times 2 =$$
$$2 \times (6.25 + (4.50 \div 2) + 2) =$$

PRACTICE

Translate each problem into math symbols. Then use the order of operations to solve each problem. You can check your answers at the end of the section.

21. Add 30 and 45. Then divide by 5.

22. Divide 81 by 9. Then multiply the quotient by 9.

23. Multiply 9 and 6. Then add 12 to the product.

24. Add the difference of 7 and 3 to the product of 2 and 8.

25. Add 12 and 4. Then multiply by 8.

26. Divide 42 by 6. Then find the difference between the quotient and 3.

27. Multiply the sum of 3 and 7 by the sum of 2 and 8.

28. Divide the sum of 15 and 5 by the product of 2 and 5

29. On Friday night, Teresa babysat for two hours. On Saturday, she babysat for four hours. On Sunday, she babysat for three and a half hours. Teresa charges $4.50 per hour of babysitting. How much money did she make babysitting over the weekend?

30. Juan and two of his friends mow lawns during the summer. They charge $25 per lawn. Altogether they made $3,000 last summer. They share the work equally and split their earnings. How many lawns did each boy mow last summer?

LESSON

2

Finding Squares and Square Roots

LESSON SUMMARY

Some math problems will ask you to calculate a square or a square root of a number. This lesson will explain what squares and square roots are and show you how to calculate them.

When you think of a square, you probably think of a box-shaped figure with four equal sides like the one shown here. As you'll see in this lesson, that's a good way to think about squares *and* square roots.

FINDING SQUARES

A *square* of a number is just the number multiplied by itself. So the square of 4 is $4 \times 4 = 16$. How does this relate to a square-shaped figure? The area of a square is the amount of space a square takes up. To calculate the area of a square, you multiply the length of one side by itself. That is why the area of a square is sometimes written as *s* squared, or s^2. Any time a number is written with a 2 raised after it, it means to multiply the number by itself, or to *square* the number.

Example: What is the square of 30?
To find the square of a number, multiply it by itself. Thus, the square of 30 is 30 × 30, or 900.

Example: Find 9^2.
When a number is followed by a raised 2, you should square it. Thus, $9^2 = 9 \times 9 = 81$.

TEST TAKING TIP

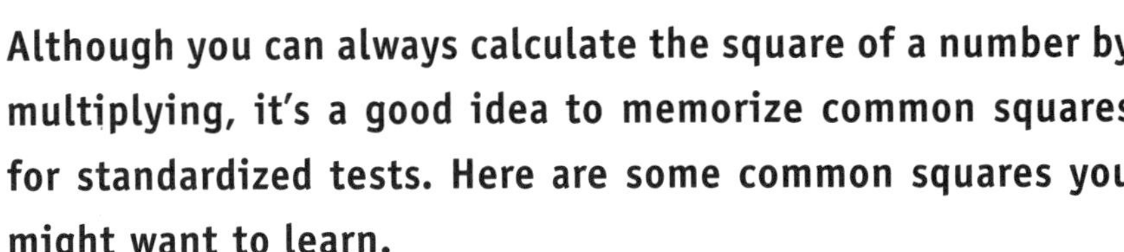

Although you can always calculate the square of a number by multiplying, it's a good idea to memorize common squares for standardized tests. Here are some common squares you might want to learn.

SQUARES TO KNOW

Number	Square	Calculation
1	1	1 × 1
2	4	2 × 2
3	9	3 × 3
4	16	4 × 4
5	25	5 × 5
6	36	6 × 6
7	49	7 × 7
8	64	8 × 8
9	81	9 × 9
10	100	10 × 10
11	121	11 × 11
12	144	12 × 12
13	169	13 × 13
14	196	14 × 14
15	225	15 × 15
16	256	16 × 16
17	289	17 × 17
18	324	18 × 18
19	361	19 × 19
20	400	20 × 20
21	441	21 × 21
22	484	22 × 22
23	529	23 × 23
24	576	24 × 24
25	625	25 × 25

PRACTICE

Solve each problem. You can check your answers at the end of the section.

1. 2^2

2. 9^2

3. 16^2

4. 12^2

5. 6^2

6. 5^2

7. 15^2

8. 8^2

9. 3^2

10. 13^2

11. 7^2

12. 26^2

13. 35^2

14. 25^2

15. 91^2

▶ FINDING SQUARE ROOTS

To find a square root of a number you have to think backwards. You will be given the area of an entire square. The answer to the problem, or *square root*, is the length of only one side of the square. That is, the square root of a number is a number that when multiplied by itself equals the number given in the problem. Keep reading. It's not as tricky as it sounds.

You may have seen this symbol before: $\sqrt{}$. This is the symbol for a square root. When it is written over a number, you are being asked to find the square root of that number.

Example: What is $\sqrt{25}$?
The problem is asking you to calculate the square root of 25. Ask yourself what number multiplied by itself equals 25. If you have memorized the list of common squares, this problem is not very hard. Even

if you haven't learned the list of common squares yet, though, you can figure this problem out: $5 \times 5 = 25$. So the square root of 25 is 5.

Example: What is the length of one side of a square that has an area of 121 square inches?

Area = 121 in.^{2}

The problem is asking you to calculate the square root of 121. Ask yourself what number multiplied by itself equals 121? You know that 11×11 is 121. Thus, the square root of 121 is 11, and the length of one side of a square with an area of 121 square inches is 11 inches.

If you aren't sure what the square root of a given square is, make a guess. Then multiply the number by itself. If it's not the correct square root, at least now you can make a better guess the second time!

PRACTICE

Solve each problem. You can check your answers at the end of the section.

16. $\sqrt{64}$

17. $\sqrt{36}$

18. $\sqrt{49}$

19. $\sqrt{81}$

20. $\sqrt{361}$

21. $\sqrt{529}$

22. $\sqrt{625}$

23. $\sqrt{256}$

24. $\sqrt{1{,}600}$

25. $\sqrt{441}$

26. $\sqrt{0}$

27. $\sqrt{3{,}600}$

28. What is the length of one side of a square that has an area of 144 square inches?

29. A square has an area of 576 square meters. What is the length of one of its sides?

30. Find the length of one of the sides of a square that has an area of 3,600 square centimeters.

In this lesson, you are working only with whole numbers. However, sometimes math problems will ask you to calculate square roots that are not whole numbers. Read the question carefully. You might be asked to round your answer to a certain place. In other cases, you might be able to use a calculator to solve the problem.

EXAMPLE: What is $\sqrt{45}$?

> **The problem is asking you what number equals 45 when multiplied by itself. You know that $6^2 = 36$ and $7^2 = 49$. Thus, the square root of 45 is a number between 6 and 7. You can find a more precise answer using a calculator.**

Real World Problems

These problems apply the skills you've learned in Section 1 to everyday situations. As you work through these problems, you'll see that the skills you've learned in this section aren't only important for math tests. They are important skills for ordinary questions that come up every day. You can check your answers at the end of the section.

Some word problems contain information that is not needed to solve the problem. Read the question carefully. Then focus only on the information you need to answer that question. Sometimes the information you need is written out in words, rather than in numerals.

1. Phat needs to read a 168-page book for school. He has four weeks to read the book. He wants to read an equal number of pages each week. How many pages should he plan to read per week?
2. Maria received $95 in cash for Christmas. She wants to buy a new backpack and a CD of her favorite musical group. The backpack she likes the most costs $85, including taxes. A similar backpack is on sale for $79, including taxes. The CD she wants to buy costs $15, also including taxes. Which backpack should Maria buy if she is set on the CD?

3. As a member of the Marketing Club at school, Aidan sold two clocks for $25 each and three clocks for $32 each. He earns $5 for his club for each sale. How much will the club receive from Aidan's sales?

4. Candy is organizing an end-of-the-year party for her class. She wants to have a swimming party. In order to have a swimming party, she must hire one lifeguard for every 20 classmates. She thinks about 60 people will come to the party. Each lifeguard charges $8 per hour. If the party is scheduled to last for three hours, how much will it cost to hire the lifeguards for the party?

5. Angela and Jon have been given a square of 64 square feet to decorate for the upcoming pep rally. Angela wants to outline their space with a gold fringe. What length of gold fringe is needed?

SECTION

1

Answers & Explanations

LESSON 1

1. $10 - 6 = 4$
2. $45 + 3 = 48$
3. $7 \times 8 = 56$
4. $20 + 3 = 23$
5. $3 + 7 = 10$
6. $63 + 2 = 65$
7. $6 \times 7 - 1 = 41$
8. $40 + 12 - 6 = 46$
9. $12 \times 6 = 72$
10. $24 \div 6 = 4$
11. $35 + 4 = 39$
12. $12 + 8 - 40 \div 10 = 12 + 8 - 4 = 16$
13. $27 - 3 = 24$
14. $14 - 1 - 2 = 11$
15. $12 + 4 + 7 = 23$
16. $15 \div 3 + 2 = 5 + 2 = 7$
17. $11 + 5 + 12 + 7 = 35$
18. $48 + 5 = 53$
19. $40 - 7 + 17 + 14 = 64$
20. $32 + 7 - 7 + 9 - 1 = 40$
21. $(30 + 45) \div 5 = 15$
22. $81 \div 9 \times 9 = 81$

23. $9 \times 6 + 12 = 66$
24. $7 - 3 + 2 \times 8 = 20$
25. $(12 + 4) \times 8 = 128$
26. $42 \div 6 - 3 = 4$
27. $(3 + 7) \times (2 + 8) = 100$
28. $(15 + 5) \div (2 \times 5) = 2$
29. First, calculate the number of hours Teresa babysat over the weekend. She babysat for 9.5 hours (2 + 4 + 3.5). Since she charges $4.50 per hour for babysitting, multiply the number of hours she babysat by $4.50. One way to set up the problem is as follows: (2 + 4 + 3.5) × $4.5 = $42.74. Thus, Teresa made $42.74 babysitting over the weekend.
30. First, determine how many lawns the three boys mowed altogether ($3,000 ÷ $25 = 120 lawns). Then, divide the total number of lawns mowed by 3. One way to set up the problem is as follows: $3,000 ÷ $25 ÷ 3 = 40 lawns. So, Juan mowed 40 lawns last summer.

LESSON 2

1. $2 \times 2 = 4$
2. $9 \times 9 = 81$
3. $16 \times 16 = 256$
4. $12 \times 12 = 144$
5. $6 \times 6 = 36$
6. $5 \times 5 = 25$
7. $15 \times 15 = 225$
8. $8 \times 8 = 64$
9. $3 \times 3 = 9$
10. $13 \times 13 = 169$
11. $7 \times 7 = 49$
12. $26 \times 26 = 676$
13. $35 \times 35 = 1{,}225$
14. $25 \times 25 = 625$
15. $91 \times 91 = 8{,}281$
16. $\sqrt{64} = 8$
17. $\sqrt{36} = 6$
18. $\sqrt{49} = 7$
19. $\sqrt{81} = 9$
20. $\sqrt{361} = 19$
21. $\sqrt{529} = 23$
22. $\sqrt{625} = 25$
23. $\sqrt{256} = 16$
24. $\sqrt{1{,}600} = 40$

25. $\sqrt{441} = 21$

26. $\sqrt{0} = 0$

27. $\sqrt{3{,}600} = 60$

28. The problem is asking you to calculate the square root of 144. Ask yourself what number multiplied by itself equals 144. You know that 12 × 12 is 144. Thus, the square root of 144 is 12, and the length of one side of a square with an area of 144 square inches is 12 inches.

29. The problem is asking you to calculate the square root of 576. The square root of 576 is 24, and the length of one side of a square with an area of 576 square meters is 24 meters.

30. The problem is asking you to calculate the square root of 3,600. The square root of 3,600 is 60, and the length of one side of a square with an area of 3,600 square centimeters is 60 centimeters.

REAL WORLD PROBLEMS

1. Divide the number of pages Phat needs to read by the number of weeks he has to read them. One way to set up the problem is as follows: 168 pages ÷ 4 weeks = 42 pages per week. Thus, Phat should read 42 pages per week.

2. First, calculate how much each purchase costs. To buy the $85 backpack, Maria would need $100 ($85 + $15). Thus, Maria should buy the $79 backpack because she doesn't have enough money to buy both the $85 backpack and the CD.

3. Aidan made five sales. One way to set up the problem is as follows: 5 × (2 + 3) = $25. Thus, the club will receive $25 from Aidan's sales. The selling prices of the clocks are not relevant to the problem.

4. First, calculate the number of lifeguards needed for the party. Then, calculate the cost of the lifeguards. One way to set up the problem is as follows: 60 ÷ 20 × 3 hours × $8 per hour = $72. Thus, the lifeguards will cost $72.

5. First, determine the length of one side of the square. The square root of 64 is 8, and the length of one side of a square with an area of 64 square feet is 8 feet. Then, calculate the perimeter of the square: 8 feet × four sides of a square = 32 feet. Thus, they will need 32 feet of gold fringe.

SECTION

2

Fractions

FRACTIONS ARE INCLUDED in any type of math question you can think of: They're in proportion questions, geometry questions, chart questions, mean questions, ratio questions, and so on. You'll learn how to think about fractions and you'll come to understand how they are used in your everyday life. You'll build up your competence dealing with fractions—no matter where they turn up.

Converting Fractions

LESSON SUMMARY

In Section 1, you learned some ways to work with whole numbers. In this lesson, you will begin working with fractions—numbers that represent parts of whole numbers. You will learn what fractions are and how to write them. You will also learn how to reduce a fraction to lowest terms and how to convert a fraction to higher terms. Lastly, you will use these skills to compare fractions.

What exactly is a fraction? Imagine that you and a friend order a whole pizza for yourselves. The pizza is cut into nine slices.

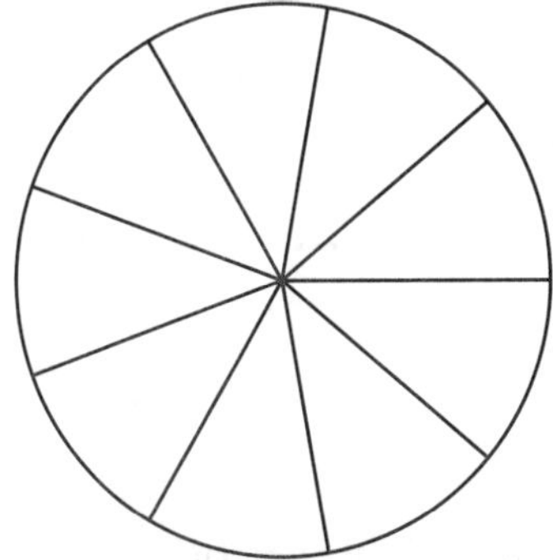

If one of you eats the whole pizza and doesn't share with the other one, then you would eat nine of the nine slices, or $\frac{9}{9}$. But what if you ate two slices and your friend ate three slices? Then you ate $\frac{2}{9}$

of the pizza, your friend ate $\frac{3}{9}$ of the pizza, and $\frac{4}{9}$ of the pizza is left over. The numbers $\frac{2}{9}$, $\frac{3}{9}$, and $\frac{4}{9}$ are all fractions.

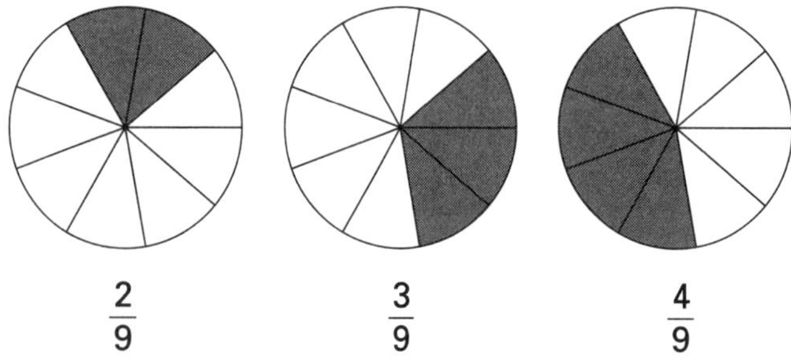

▶ WRITING AND RECOGNIZING FRACTIONS

Notice that *fractions* are two numbers that represent a part of a whole. The two numbers are separated by a bar. The bar means "divide the top number by the bottom number."

The top number is called the *numerator*. The numerator tells you how many parts of the whole are being talked about. For example, $\frac{2}{9}$ of the pizza shown above refers to two slices of a pizza that has been cut into nine slices.

The bottom number in a fraction is called the *denominator*. The denominator tells you how many equal parts the whole has been divided into. The pizza shown previously has been divided into nine slices, so the denominator is 9. What if you had cut the pizza into eight slices? Then the denominator would be 8.

You can keep the numerator and the denominator straight by remembering that both *denominator* and *down* begin with the letter *d*. The denominator is *down* below the bar.

Think about some other common fractions.

- ▶ We use fractions to talk about money. For example, a quarter is 25 cents, or $\frac{1}{4}$ of a dollar. Four quarters, or $\frac{1}{4}$, equal one dollar.
- ▶ We also use fractions to talk about time. An hour is a fraction of a day. One hour is $\frac{1}{24}$ of a whole day. One day is a fraction of a week: $\frac{1}{7}$. What fraction of a year is one month?
- ▶ Your school grades are probably written in fractions. If you receive a 90 out of a total of 100 possible points, then your grade is the fraction $\frac{90}{100}$. Some teachers grade out of 20 possible points. If you receive a 19 out of 20 points, then your grade is the fraction $\frac{19}{20}$.

- Sometimes you will see fractions in ads. A department store might have a sale of one-half off last season's styles. That means that you pay only $\frac{1}{2}$ of the whole price.

Example: Write a fraction to represent the part of the whole that is shaded in.

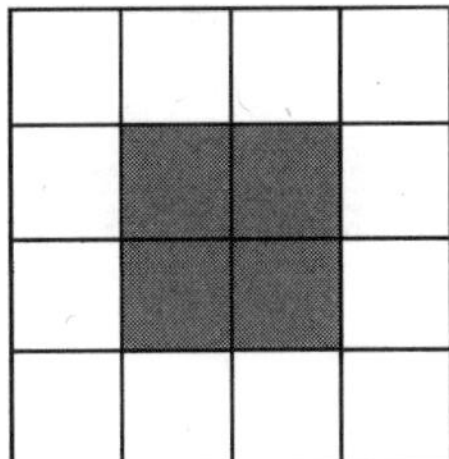

To solve this problem, you first count the number of equal parts the whole is divided into. The figure is divided into 16 squares of equal size, so the denominator of the fraction is 16. Then count the number of the parts that are shaded in: 4. So the fraction of the whole that is shaded in is $\frac{4}{16}$.

A pizza is one thing. But sometimes a *whole* might refer to a group of things. For example, a six-pack of soft drinks is six things, but only one whole. Think of some other examples of wholes that aren't single objects.

- 25 desks in a classroom
- 20 basketballs in your school gym
- 12 donuts in a package of donuts
- 300 students in your grade

Example: Your teacher assigned ten homework grades this semester. But you turned in only eight of the assignments. What fraction of the homework assignments did you turn in? What fraction of the homework assignments did you fail to turn in?

First, determine the denominator. There are 10 homework assignments, so the denominator of both fractions is 10. You turned in 8 homework assignments, so you turned in $\frac{8}{10}$ of the assignments. You didn't turn in 2 of the homework assignments, so you are missing $\frac{2}{10}$ of the assignments.

PRACTICE

Write a fraction to represent the part of each whole that is shaded in. You can check your answers at the end of the section.

1.

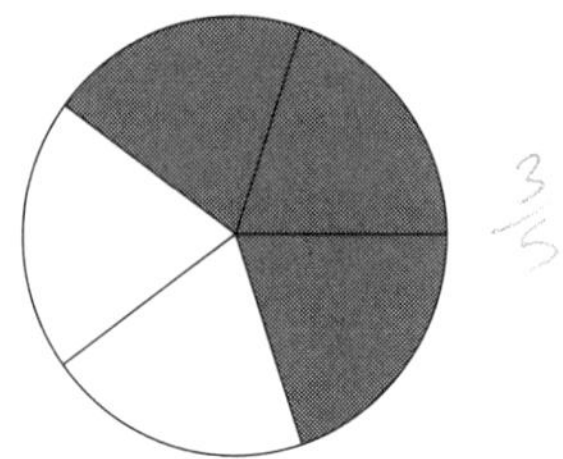

2.

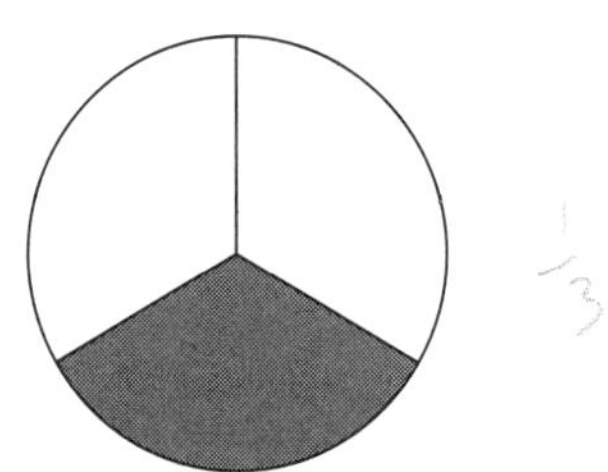

3.

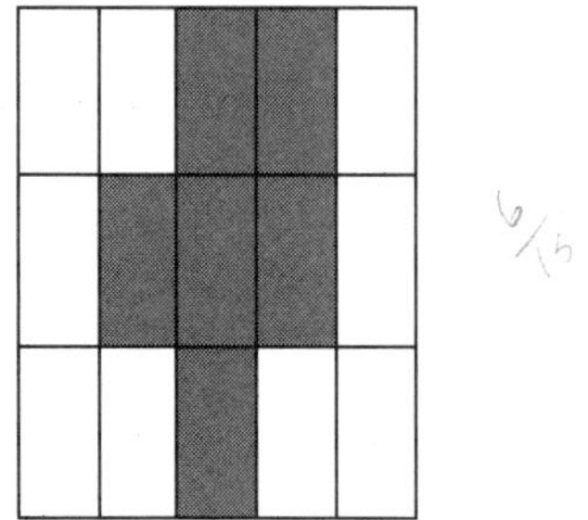

4.

5.

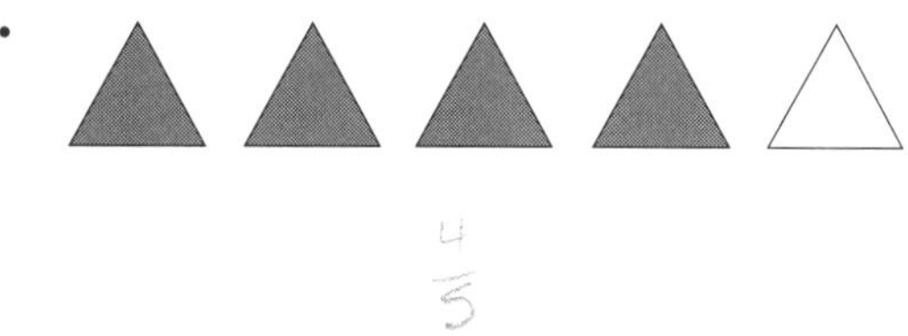

6.

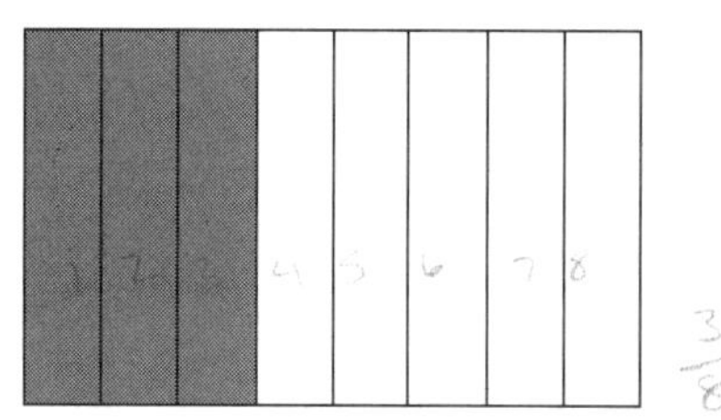

7.

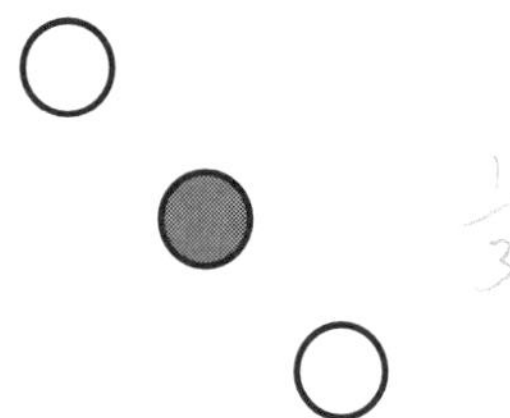

Solve each problem.

8. Out of twenty students in Karen's French class, three have visited France. What fraction of the students in Karen's French class has visited France?

9. The school cafeteria serves hamburgers every Friday. What fraction of each week does the cafeteria serve hamburgers?

10. A store that rents videos stocks 900 videos. Only 40 of these videos are rated G. What fraction of the store's videos is rated G?

REDUCING FRACTIONS TO LOWEST TERMS

The numerator and the denominator of a fraction are called *terms*. Often the directions on math tests will tell you to *reduce fractions to lowest terms*. This is the standard and usual way to write fractions. Even if the directions do not tell you to reduce to lowest terms, you should always reduce your fractions to lowest terms.

To reduce a fraction to lowest terms, you need to find a fraction that is equal to the one you have and write it with a smaller numerator and denominator. Divide both the numerator and the denominator by the same whole number. The whole number must divide evenly into both numbers. Continue to divide the numerator and denominator until there is no number other than 1 that can divide evenly into both numbers. If only 1 divides into both numbers evenly, then the fraction is said to be *reduced to lowest terms.*

Example: Reduce $\frac{7}{28}$ to lowest terms.
Begin by thinking of a number that will divide evenly into both the numerator and the denominator. You know that 4 × 7 = 28, so you can divide both numbers by 7.

$$\frac{7 \div 7}{28 \div 7} = \frac{1}{4}$$

The fraction $\frac{1}{4}$ is equal to $\frac{7}{28}$. No number other than 1 can divide into both 1 and 4 ; so $\frac{1}{4}$ is reduced to lowest terms.

If you don't know where to begin to reduce a fraction, look at the last digits of the numerator and denominator.

- **If both the numerator and the denominator end in zero, then cross out the same number of zeros in each number. If you cross out one zero in each number, you are dividing by 10. Crossing out two zeros at the end of each number is dividing by 100.**

EXAMPLE: Reduce $\frac{300}{7,000}$ to lowest terms.
Both 300 and 7,000 end in zeros. You can cross out two zeros in each number.

$$\frac{3\not{0}\not{0}}{7,0\not{0}\not{0}} = \frac{3}{70}$$

- **If both numbers are even, but don't end in zero, you can begin by dividing by 2.**

continued from previous page

EXAMPLE: Reduce $\frac{12}{22}$ to lowest terms.

Both numbers are even. Begin by dividing by 2:

$$\frac{12 \div 2}{22 \div 2} = \frac{6}{11}$$

- Often when you begin reducing a fraction by dividing by a small number, you will have to divide more than one time. If you can find the largest number that divides into both numbers, you can reduce the fraction faster—often in only one step.

EXAMPLE: Reduce $\frac{8}{24}$ to lowest terms.

If you begin by dividing by 2, you will have to divide again by 4:

First, divide by 2:

$$\frac{8 \div 2}{24 \div 2} = \frac{4}{12}$$

Then divide by 4:

$$\frac{4 \div 4}{12 \div 4} = \frac{1}{3}$$

If you begin by dividing by 8, you can reduce the fraction with only one step:

$$\frac{8 \div 8}{24 \div 8} = \frac{1}{3}$$

Notice that you get the same answer either way.

PRACTICE

Reduce each fraction to lowest terms. You can check your answers at the end of the section.

11. $\frac{12}{48}$

12. $\frac{7}{49}$

13. $\frac{9}{81}$

14. $\frac{42}{88}$

15. $\frac{150}{250}$

16. $\frac{240}{360}$

17. $\frac{18}{27}$

Write the answer to each problem in lowest terms.

18. Jean spent two hours working on her history project last Saturday. If she spent a total of ten hours on the project, what fraction of the total time did she spend on Saturday?

19. Patrick spent $40 of his savings over the weekend. If he had saved $100, what fraction of his savings did he spend?

20. Chris made two pizzas for his family. He cut each pizza into eight slices. Three slices of pizza were left over. What fraction of the pizzas did the family eat?

▶ RAISING FRACTIONS TO HIGHER TERMS

You raise a fraction to higher terms by multiplying both the numerator and the denominator by the same number. For example, if you multiply $\frac{5}{6}$ by $\frac{4}{4}$, you will get $\frac{20}{24}$.

$$\frac{5}{6} \times \frac{4}{4} = \frac{20}{24}$$

Both $\frac{5}{6}$ and $\frac{20}{24}$ are equal fractions. You sometimes need to raise a fraction to higher terms when you are comparing fractions. You will also have to raise fractions to higher terms when you are adding and subtracting fractions that have different denominators. You'll learn more about adding and subtracting fractions in the next lesson.

When raising a fraction to higher terms, you will usually need to find an equal fraction with a specific denominator. Here's how to raise a fraction to higher terms with a specific denominator.

Step 1: Divide the denominator of the fraction into the new denominator.

Step 2: Multiply the quotient, or the answer to Step 1, by the numerator.

Step 3: Write the product, or the answer to Step 2, over the new denominator.

Example: $\frac{1}{3} = \frac{?}{9}$

This problem asks you to raise $\frac{1}{3}$ to 9ths.

Step 1: Divide the denominator into the new denominator. The new denominator is 9.

$9 \div 3 = 3$

Step 2: Multiply the answer to Step 1 by the numerator.

$1 \times 3 = 3$

Step 3: Write the answer to Step 2 over the new denominator.

$\frac{3}{9}$

PRACTICE

Raise each fraction to higher terms with the denominator given. You can check your answers at the end of the section.

21. $\frac{2}{3} = \frac{?}{12}$

22. $\frac{5}{8} = \frac{?}{32}$

23. $\frac{3}{8} = \frac{?}{32}$

24. $\frac{12}{20} = \frac{?}{40}$

25. $\frac{9}{10} = \frac{?}{50}$

26. $\frac{7}{13} = \frac{?}{39}$

27. $\frac{3}{10} = \frac{?}{120}$

28. $\frac{20}{50} = \frac{?}{100}$

29. $\frac{2}{9} = \frac{?}{810}$

30. $\frac{1}{8} = \frac{?}{640}$

Sometimes you will be asked to compare two or more fractions. You might be asked which of two fractions is larger, for example. Or you could be asked if two fractions are equal. If the two fractions have the same denominator, you simply compare the numerators. The number with the higher numerator is larger.

EXAMPLE: Which fraction is larger, $\frac{5}{12}$ or $\frac{8}{12}$?

You know that 8 is larger than 5, so $\frac{8}{12}$ is larger than $\frac{5}{12}$. What if the fractions do not have the same denominator? One way to answer this type of question is to convert both fractions so that they have the same denominator. This is called finding a *common denominator.* If the two fractions have the same denominator, you can compare their numerators.

EXAMPLE: Which fraction is larger, $\frac{5}{6}$ or $\frac{3}{4}$?

To answer this question, you can raise both fractions to higher terms with a common denominator. First, you need to find a denominator that both 6 and 4 can divide into. You could choose 24 ($6 \times 4 = 24$), or 12. Then raise each fraction to higher terms with the same denominator.

$$\frac{5}{6} \times \frac{2}{2} = \frac{10}{12}$$

$$\frac{3}{4} \times \frac{3}{3} = \frac{9}{12}$$

Then compare the numerators of the two fractions. Because 10 is greater than 9, you know that $\frac{10}{12}$ is larger than $\frac{9}{12}$.

Adding and Subtracting Fractions

LESSON SUMMARY

In this lesson, you will learn how to add and subtract fractions.

How would you add 2 hours and \$5? You can't. You can add and subtract only like objects. You can add \$2 to \$5 or 2 hours to 5 hours. It's the same with fractions. To add and subtract fractions, you need *like fractions.* Like fractions are fractions that have the same denominator.

ADDING AND SUBTRACTING LIKE FRACTIONS

If fractions have the same denominator, you add and subtract their numerators. Then write the sum or difference over the denominator.

Example: $\frac{1}{12} + \frac{2}{12} + \frac{1}{12}$
All three of these fractions have the denominator 12. Thus, they are like fractions.

Step 1: Add the numerators together.

$1 + 2 + 1 = 4$

Step 2: Write the sum of the numerators over the denominator.

$\frac{4}{12}$

Step 3: Reduce your answer to lowest terms.

$\frac{4 \div 4}{12 \div 4} = \frac{1}{3}$

So, $\frac{1}{12} + \frac{2}{12} + \frac{1}{12} = \frac{1}{3}$.

Example: $\frac{5}{8} - \frac{3}{8}$

Both fractions have the denominator 8.

Step 1: Subtract the numerators.

$5 - 3 = 2$

Step 2: Write the difference over the common denominator.

$\frac{2}{8}$

Step 3: Reduce your answer to lowest terms.

$\frac{2 \div 2}{8 \div 2} = \frac{1}{4}$

So, $\frac{5}{8} - \frac{3}{8} = \frac{1}{4}$.

When subtracting fractions, the order of the fractions is important. Write the numerator that you are subtracting *from* first. Then subtract as you would any two numbers.

PRACTICE

Write the answer to each problem in lowest terms. You can check your answers at the end of the section.

1. $\frac{1}{5} + \frac{3}{5} =$

2. $\frac{1}{3} + \frac{1}{3} =$

3. $\frac{2}{6} + \frac{3}{6} =$

4. $\frac{7}{15} + \frac{1}{15} + \frac{2}{15} =$

5. $\frac{1}{20} + \frac{6}{20} + \frac{3}{20} =$

6. $\frac{8}{9} - \frac{2}{9} =$

7. $\frac{7}{10} - \frac{1}{10} =$

8. $\frac{3}{11} - \frac{1}{11} =$

9. $\frac{12}{25} - \frac{2}{25} =$

10. $\frac{4}{15} - \frac{1}{15} =$

▶ ADDING AND SUBTRACTING UNLIKE FRACTIONS

Fractions that have different denominators are called *unlike fractions.* Before you can add or subtract unlike fractions, you first need to change the fractions so that they have the same number in the denominator. This process is called *finding a common denominator.*

There are two main ways to find a common denominator. One way is to multiply each denominator by 2, 3, 4, 5, and so on. Then compare the lists of multiples of each denominator. The numbers that are the same, or that are in common, are common denominators.

Example: Find a common denominator for $\frac{1}{4}$ and $\frac{1}{6}$.
List multiples for each denominator.
Multiples of 4: 4, 8, 12, 16
Multiples of 6: 6, 12, 18, 24
The numbers 4 and 6 share the multiple 12. So, 12 is a common denominator for $\frac{1}{4}$ and $\frac{1}{6}$.

Another way to find a common denominator is to multiply the denominators together.

Example: Find a common denominator for $\frac{1}{4}$ and $\frac{1}{6}$.
Multiply the denominators together: $4 \times 6 = 24$. So, 24 is a common denominator for $\frac{1}{4}$ and $\frac{1}{6}$.

Using the smallest common denominator, called the *lowest common denominator*, will make your calculations easier. So, in the case of $\frac{1}{4}$ and $\frac{1}{6}$, 12 is the lowest common denominator. Although using 24 as a common denominator will work just fine for your calculations, your numbers will be larger and you will have to reduce your final answer to lowest terms.

Follow these steps when adding or subtracting unlike fractions.

Step 1: Find a common denominator.

Step 2: Change each fraction so that it has that common denominator.

Step 3: Add or subtract the fractions as indicated.

Step 4: Reduce your answer to lowest terms.

Example: $\frac{1}{4} + \frac{1}{6}$

Step 1: You already know that the lowest common denominator is 12.

Step 2: Raise each fraction to higher terms with the denominator 12.

$$\frac{1 \times 3}{4 \times 3} = \frac{3}{12}$$

$$\frac{1 \times 2}{6 \times 2} = \frac{2}{12}$$

Step 3: Add the fractions. The fractions are now like fractions, so you add the numerators and write the sum over the common denominator.

$$\frac{3}{12} + \frac{2}{12} = \frac{5}{12}$$

Step 4: Reduce your answer to lowest terms. Both the numerator and the denominator can only be divided by 1, so your answer is already in lowest terms.

What if you had chosen 24 as your common denominator in the last example instead of 12? Let's redo the problem with 24 as the common denominator.

Step 1: You decide to use 24 as the common denominator.
Step 2: Raise each fraction to higher terms with the denominator 24.

$$\frac{1 \times 6}{4 \times 6} = \frac{6}{24}$$

$$\frac{1 \times 4}{6 \times 4} = \frac{4}{24}$$

Step 3: Add the fractions.

$$\frac{6}{24} + \frac{4}{24} = \frac{10}{24}$$

Step 4: Reduce your answer to lowest terms. Both the numerator and the denominator can be divided by 2.

$$\frac{10 \div 2}{24 \div 2} = \frac{5}{12}$$

So, the answer is the same: $\frac{5}{12}$.

Example: $\frac{1}{2} - \frac{1}{3}$

Step 1: Find the lowest common denominator: $2 \times 3 = 6$
Step 2: Raise each fraction to higher terms with the denominator 6.

$$\frac{1 \times 3}{2 \times 3} = \frac{3}{6}$$

$$\frac{1 \times 2}{3 \times 2} = \frac{2}{6}$$

Step 3: Subtract the fractions. The fractions are now like fractions, so you subtract the numerators and write the difference over the common denominator.

$$\frac{3}{6} - \frac{2}{6} = \frac{1}{6}$$

Step 4: Reduce your answer to lowest terms. Both the numerator and the denominator can only be divided by 1, so your answer is already in lowest terms.

PRACTICE

Write the answer to each problem in lowest terms. You can check your answers at the end of the section.

11. $\frac{1}{5} + \frac{6}{10} =$

12. $\frac{2}{7} + \frac{3}{21} =$

13. $\frac{2}{15} + \frac{5}{20} =$

14. $\frac{1}{5} + \frac{6}{12} =$

15. $\frac{1}{2} + \frac{1}{4} + \frac{1}{6} =$

16. $\frac{3}{8} + \frac{1}{4} + \frac{3}{16} =$

17. $\frac{3}{4} - \frac{1}{2} =$

18. $\frac{5}{6} - \frac{1}{2} =$

19. $\frac{1}{3} - \frac{1}{9} =$

20. $\frac{1}{4} - \frac{1}{5} =$

21. $\frac{1}{2} - \frac{1}{9} =$

22. $\frac{3}{4} - \frac{6}{20} =$

23. $\frac{7}{8} - \frac{3}{9} =$

24. $\frac{7}{16} - \frac{1}{4} =$

25. $\frac{9}{10} - \frac{12}{15} =$

You know from the pizza example in Lesson 3 that $\frac{9}{9} = 1$. In fact, any number over itself is equal to 1.

$\frac{1}{1} = 1$ $\frac{2}{2} = 1$ $\frac{3}{3} = 1$

$\frac{4}{4} = 1$ $\frac{5}{5} = 1$

and so on.

The closer the number in the numerator is to the number in the denominator, the closer the fraction is to 1.

Multiplying and Dividing Fractions

LESSON SUMMARY

In this lesson, you will learn how to multiply and divide fractions.

You do not need to find a common denominator when multiplying or dividing fractions. In this sense, multiplying and dividing fractions is easier than adding and subtracting them. If you know how to multiply, then you basically already know how to multiply and divide fractions.

MULTIPLYING FRACTIONS

Follow these steps when multiplying fractions.

Step 1: Multiply the numerators. Write the product as the numerator of your answer.

Step 2: Multiply the denominators. Write the product as the denominator of your answer.

Step 3: Reduce your answer to lowest terms.

Example: $\frac{2}{3} \times \frac{1}{2}$

Step 1: Multiply the numerators. Write the product as the numerator of your answer.

$2 \times 1 = 2$

So the top number of your answer is 2.

Step 2: Multiply the denominators. Write the product as the denominator of your answer.

$3 \times 2 = 6$

So your fraction is $\frac{2}{6}$.

Step 3: Reduce your answer to lowest terms.
Both 2 and 6 can be divided by 2:

$\frac{2 \div 2}{6 \div 2} = \frac{1}{3}$

So your answer written in lowest terms is $\frac{1}{3}$.

When a fraction is followed by the word *of* it means to multiply.

EXAMPLE: What is $\frac{5}{8}$ of $\frac{1}{2}$?
The problem is asking you to multiply $\frac{5}{8}$ and $\frac{1}{2}$.

Step 1: Multiply the numerators. Write the product as the numerator of your answer.

$5 \times 1 = 5$

So the top number of your answer is 5.

Step 2: Multiply the denominators. Write the product as the denominator of your answer.

$8 \times 2 = 16$

So your fraction is $\frac{5}{16}$.

Step 3: The numbers 5 and 16 can both be divided only by 1. So your answer written in lowest terms is $\frac{5}{16}$.

Example: Julie bought $\frac{3}{4}$ yard of fabric to make a dress for her baby cousin. She used only $\frac{2}{3}$ of the piece of fabric. How much of the fabric did she use for the dress?
The problem is asking you to multiply $\frac{3}{4}$ and $\frac{2}{3}$.

Step 1: Multiply the numerators. Write the product as the numerator of your answer.

$3 \times 2 = 6$

So the top number of your answer is 6.

Step 2: Multiply the denominators. Write the product as the denominator of your answer.

$4 \times 3 = 12$

So your fraction is $\frac{6}{12}$.

Step 3: Reduce your answer to lowest terms.
Both 6 and 12 can be divided by 6:

$\frac{6 \div 6}{12 \div 6} = \frac{1}{2}$

So your answer written in lowest terms is $\frac{1}{2}$ yard of fabric.

You can simplify your multiplication by *canceling* before multiplying. Like reducing a fraction, canceling involves dividing. If you can see a number that will divide evenly into one of the numerators and one of the denominators of each fraction you are multiplying, then do so. This is canceling.

EXAMPLE: Let's look at the last problem again and try to solve it by canceling:

$\frac{3}{4} \times \frac{2}{3}$

There is a 3 in the numerator of the first fraction and in the denominator of the second fraction. You can cancel by dividing by 3:

$\frac{\overset{1}{\cancel{3}}}{4} \times \frac{2}{\underset{1}{\cancel{3}}}$

You have simplified your problem to $\frac{1}{4} \times \frac{2}{1}$. Do you see a way to further simplify this problem with canceling? Both 4 and 2 can be divided by 2:

$\frac{1}{\underset{2}{\cancel{4}}} \times \frac{\overset{1}{\cancel{2}}}{1}$

So now your multiplication is very easy: $\frac{1}{2} \times \frac{1}{1} = \frac{1}{2}$.

PRACTICE

Write the answer to each problem in lowest terms. You can check your answers at the end of the section.

1. $\frac{1}{3} \times \frac{1}{2} =$
2. $\frac{2}{5} \times \frac{1}{5} =$
3. $\frac{3}{12} \times \frac{4}{20} =$
4. $\frac{6}{7} \times \frac{3}{4} =$
5. $\frac{2}{20} \times \frac{6}{8} =$
6. $\frac{7}{21} \times \frac{3}{7} =$
7. $\frac{3}{100} \times \frac{10}{17} =$
8. $\frac{3}{7} \times \frac{7}{9} =$
9. What is $\frac{6}{15}$ of $\frac{5}{8}$?
10. Find $\frac{8}{9}$ of $\frac{5}{6}$.
11. A recipe calls for $\frac{3}{4}$ a cup of sugar to make two dozen brownies. Elizabeth wants to make only one dozen brownies. How much sugar does she need?
12. Jake ran half way around his school's track. If one loop on the school's track is a quarter mile, how far did Jake run?

DIVIDING FRACTIONS

Dividing fractions is very similar to multiplying fractions. To divide a fraction by another fraction follow these steps.

Step 1: Invert the second fraction. That is, write the numerator on the bottom and the denominator on the top.

Step 2: Multiply the two fractions.

Step 3: Write the answer in lowest terms.

Example: $\frac{1}{3} \div \frac{3}{4}$

Step 1: Invert the second fraction.

$\frac{3}{4}$ inverts to $\frac{4}{3}$

Step 2: Multiply.

$\frac{1}{3} \times \frac{4}{3} = \frac{4}{9}$

Step 3: The numbers 4 and 9 can both be divided evenly only by 1, so this fraction is already written in lowest terms.

The answer is $\frac{4}{9}$.

Some math problems include whole numbers and fractions. You can write a whole number as a fraction by writing the whole number over 1.

$1 = \frac{1}{1}$

$2 = \frac{2}{1}$

$3 = \frac{3}{1}$

$4 = \frac{4}{1}$

$5 = \frac{5}{1}$ **and so on**

EXAMPLE: $\frac{1}{5} \div 10 =$

Rewrite the question so that the whole number is a fraction.

$\frac{1}{5} \div \frac{10}{1} =$

Step 1: Invert the second fraction.

$\frac{10}{1}$ **inverts to** $\frac{1}{10}$

Step 2: Multiply.

$\frac{1}{5} \times \frac{1}{10} = \frac{1}{50}$

Step 3: The numbers 1 and 50 can both be divided evenly only by 1, so this fraction is already written in lowest terms.

The answer is $\frac{1}{50}$.

PRACTICE

Write the answer to each problem in lowest terms. You can check your answers at the end of the section.

13. $\frac{1}{4} \div \frac{1}{3} =$

14. $\frac{1}{3} \div \frac{1}{2} =$

15. $\frac{1}{2} \div \frac{5}{6} =$

16. $\frac{3}{9} \div \frac{8}{10} =$

17. $\frac{1}{8} \div \frac{1}{3} =$

18. $\frac{1}{9} \div \frac{1}{6} =$

19. $\frac{2}{7} \div \frac{2}{5} =$

20. $\frac{1}{8} \div \frac{3}{4} =$

21. $\frac{2}{25} \div \frac{4}{5} =$

22. $\frac{1}{20} \div \frac{9}{10} =$

23. $\frac{1}{60} \div \frac{5}{12} =$

24. Jason helped make hamburgers for his family's cookout. His mother told him to use about $\frac{1}{3}$ pound of meat for each burger. How many burgers can he make if he has five pounds of meat?

25. Thuy has ten cups of brown sugar. She wants to make as many batches of cookies for a bake sale as possible. If each batch of cookies requires $\frac{2}{3}$ cup of brown sugar, how many batches of cookies can Thuy make? (Hint: When writing your final answer to this problem, remember that the bar in a fraction means to divide the top number by the bottom number.)

Working with Improper Fractions and Mixed Numbers

LESSON SUMMARY

In this lesson, you will learn the difference between a proper fraction and an improper fraction and how to convert an improper fraction to a mixed number. You will also practice adding, subtracting, multiplying, and dividing when a problem includes an improper fraction or a mixed number.

So far, the fractions you have been working with in this book have all been proper fractions. A *proper fraction* is one in which the numerator is smaller than the denominator. These are examples of proper fractions: $\frac{1}{2}$, $\frac{2}{3}$, $\frac{5}{6}$, $\frac{34}{91}$, and so on. Proper fractions are always equal to less than 1. They represent a part of whole.

WHAT ARE IMPROPER FRACTIONS?

The numerator of an *improper fraction* is greater than its denominator. Here are some examples of improper fractions: $\frac{4}{2}$, $\frac{25}{5}$, $\frac{12}{4}$, $\frac{10}{3}$, and so on. Remember that the bar in a fraction means to divide the top number by the bottom number. Now, look again at the examples of improper fractions.

Let's try dividing the first one: $\frac{4}{2}$. What is $4 \div 2$? Yes, it's 2. So the improper fraction $\frac{4}{2} = 2$. Now you try $\frac{25}{5}$. Do you see that $\frac{25}{5} = 5$? Notice the pattern here. Improper fractions are all equal to or greater than 1.

Example: Write a fraction to represent the shaded part of the figure.

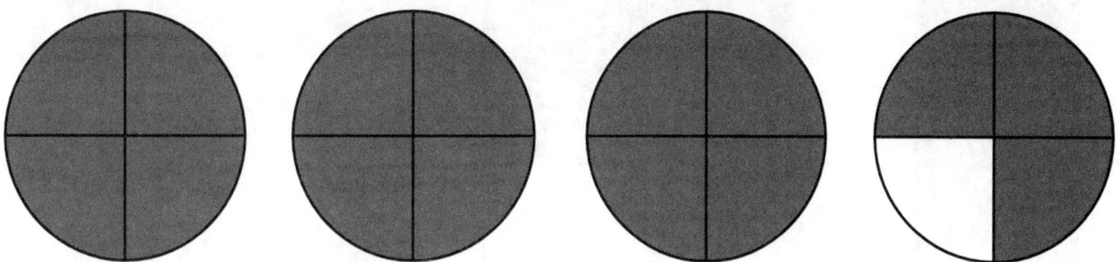

To solve this problem, you first count the number of equal parts the circles have been divided into. Each circle is divided into four wedges of equal size, so the denominator of the fraction is 4. Then count the number of the parts that are shaded in: 15. So the fraction that is shaded in is $\frac{15}{4}$. Notice that this is an improper fraction because 15 is greater than 4.

PRACTICE

Write a fraction to represent the shaded part of each figure. You can check your answers at the end of the section.

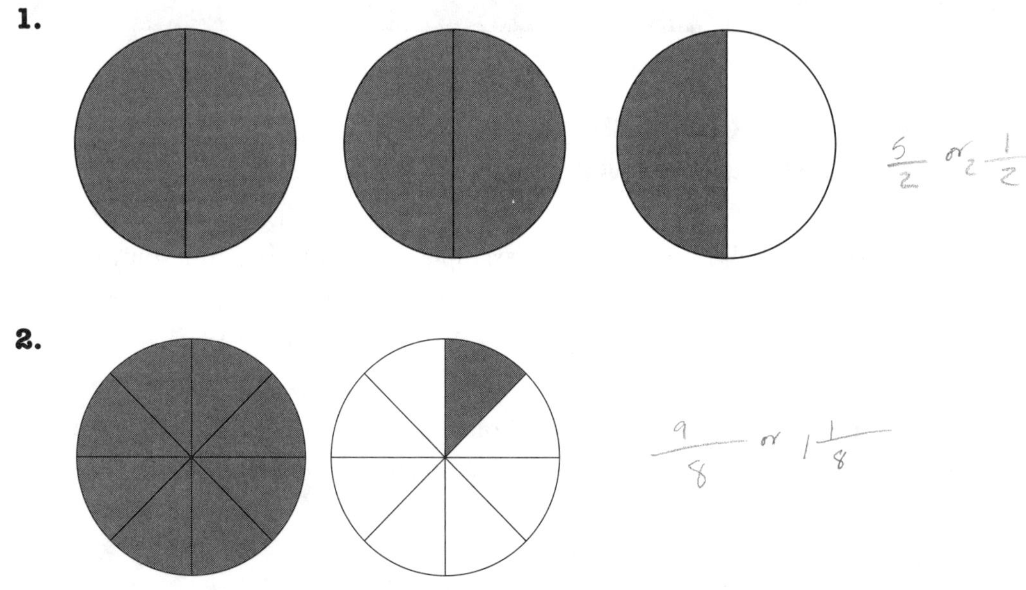

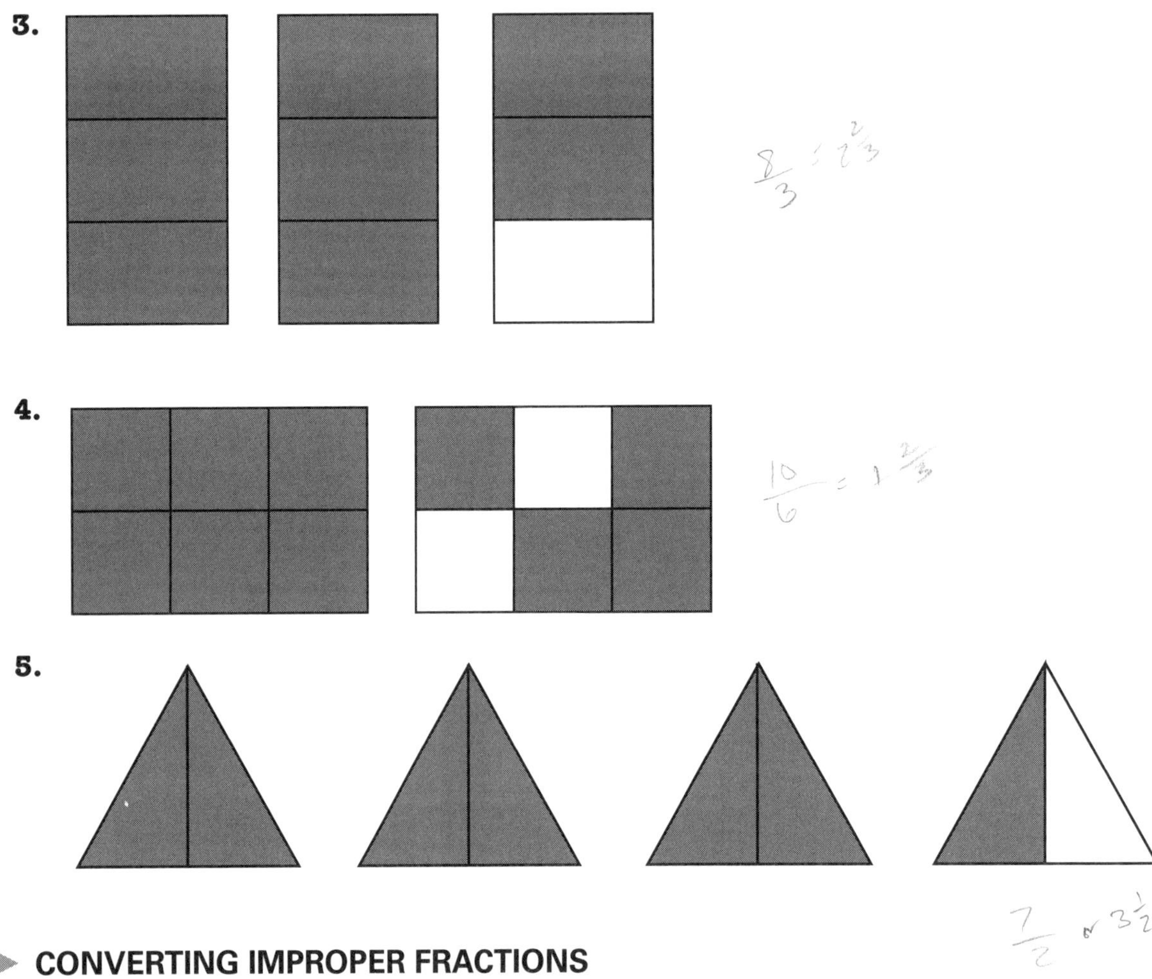

CONVERTING IMPROPER FRACTIONS TO WHOLE NUMBERS AND MIXED NUMBERS

Many improper fractions are equal to whole numbers. For example, $\frac{4}{2} = 2$ and $\frac{25}{5} = 5$. But some improper fractions are not equal to a whole number. They represent a whole number plus a proper fraction. A whole number plus a proper fraction is called a *mixed number*. Examples of mixed numbers are $1\frac{2}{3}$, $3\frac{1}{2}$, $25\frac{7}{8}$, and so on.

Improper fractions should usually be converted to whole numbers or mixed numbers when writing your final answer to a problem. Here's how to convert an improper fraction to a mixed number.

Step 1: Divide the numerator by the denominator.
Step 2: Write the number of times the denominator divides into the numerator as the whole number part of the mixed number.
Step 3: Write the remainder on top of the improper fraction's denominator.

Example: Convert $\frac{13}{3}$ to a mixed number.

Step 1: Divide the numerator by the denominator.

$3\overline{)13}$

Step 2: Write the number of times the denominator divides into the numerator as the whole number part of the mixed number.

$$\begin{array}{r} 4 \\ 3\overline{)13} \\ -12 \end{array}$$ so the mixed number will begin with the whole number 4

Step 3: Write the remainder on top of the improper fraction's denominator.

$$\begin{array}{r} 4 \\ 3\overline{)13} \\ \underline{-12} \\ 1 \end{array}$$ so you write 1 over 3, to get $\frac{1}{3}$

You keep the same denominator → $3\overline{)13}$

4 ← The whole number is the whole number part of the mixed number.

-12

1 ← The remainder becomes the new numerator.

The mixed number equal to $\frac{13}{3}$ is $4\frac{1}{3}$.

What if the improper fraction in the last example had been $\frac{12}{3}$? Let's follow the same steps.

Step 1: Divide the numerator by the denominator.

$$3\overline{)12}$$

Step 2: Write the number of times the denominator divides into the numerator as the whole number part of the mixed number.

$$\begin{array}{r} 4 \\ 3\overline{)12} \\ -12 \end{array}$$ **so the mixed number will begin with the whole number 4**

Step 3: Write the remainder on top of the improper fraction's denominator.

$$\begin{array}{r} 4 \\ 3\overline{)12} \\ \underline{-12} \\ 0 \end{array}$$ **if there is no remainder, then the answer is a whole number**

The whole number equal to $\frac{12}{3}$ is 4.

When there is no remainder, the answer will be a whole number rather than a mixed number.

PRACTICE

Convert each improper fraction to a mixed number or a whole number. You can check your answers at the end of the section.

6. $\frac{7}{7}$

7. $\frac{15}{3}$

8. $\frac{5}{4}$

9. $\frac{36}{6}$

10. $\frac{7}{5}$

11. $\frac{29}{4}$

12. $\frac{50}{7}$

13. $\frac{500}{50}$

14. $\frac{14}{4}$

15. $\frac{65}{60}$

▶ CONVERTING WHOLE NUMBERS AND MIXED NUMBERS TO IMPROPER FRACTIONS

When multiplying a whole number or a mixed number with a fraction, you will need to convert the whole number or mixed number to an improper fraction. You already know how to change a whole number into an improper fraction: you just write the whole number over 1.

Example: Convert 12 to an improper fraction.
Write the whole number over 1: $\frac{12}{1}$.

Here's how to convert a mixed number to an improper fraction.

Step 1: Multiply the whole number part of the mixed number by the denominator of the fraction.
Step 2: Add the numerator of the fraction to the product (the answer to step 1).
Step 3: Write the sum (the answer to step 2) as the numerator in the improper fraction.
Step 4: Keep the same denominator.

Example: Convert $2\frac{2}{3}$ to an improper fraction.
Step 1: Multiply the whole number part of the mixed number by the denominator of the fraction.

$$2 \times 3 = 6$$

Step 2: Add the numerator of the fraction to the product (the answer to step 1).

$6 + 2 = 8$

Step 3: Write the sum (the answer to step 2) as the numerator in the improper fraction.

So the numerator in the new fraction will be 8.

Step 4: Keep the same denominator.

So the denominator in the new fraction will be 3.

The answer is $\frac{8}{3}$.

PRACTICE

Convert each number to an improper fraction. You can check your answers at the end of the section.

16. 17

17. $6\frac{1}{2}$

18. $3\frac{3}{4}$

19. $5\frac{1}{6}$

20. $7\frac{4}{5}$

21. $12\frac{2}{3}$

22. $8\frac{6}{7}$

23. $1\frac{3}{50}$

24. $2\frac{23}{40}$

25. $6\frac{4}{10}$

▶ ADDING AND SUBTRACTING MIXED NUMBERS

When adding and subtracting mixed numbers, you add the fractions and the whole numbers separately and then put them together at the end. Follow these steps to add mixed numbers.

Step 1: If the fractions do not have a common denominator, change them into fractions with a common denominator.
Step 2: Add the numerators of the fractions. Write the sum over the common denominator.
Step 3: Add the whole numbers.

Step 4: If your answer includes an improper fraction, convert the improper fraction to a mixed number and add it to the whole number part of your answer.

Example: $3\frac{1}{5} + 6\frac{3}{5} =$

Step 1: If the fractions do not have a common denominator, change them into fractions with a common denominator.

Both $\frac{1}{5}$ and $\frac{3}{5}$ have a 5 in the denominator, so you do not need to do anything.

Step 2: Add the numerators of the fractions. Write the sum over the common denominator.

$1 + 3 = 4$, so 4 is the numerator of your new fraction.
The new fraction is $\frac{4}{5}$.

Step 3: Add the whole numbers.

$3 + 6 = 9$

Step 4: Write the whole number and the fraction beside one another.

$9\frac{4}{5}$

Step 5: If your answer includes an improper fraction, convert the improper fraction to a mixed number and add it to the whole number part of your answer.

Because 4 is less than 5, $\frac{4}{5}$ is a proper fraction. You do not need to do anything. The answer is $9\frac{4}{5}$.

Example: $1\frac{4}{5} + 7\frac{3}{4} =$

Step 1: If the fractions do not have a common denominator, change them into fractions with a common denominator.

Both $\frac{4}{5}$ and $\frac{3}{4}$ have different denominators, so you need to find a common denominator. You might choose to use 20 as a common denominator since both 4 and 5 divide evenly into 20.

$\frac{4 \times 4}{5 \times 4} = \frac{16}{20}$

$\frac{3 \times 5}{4 \times 5} = \frac{15}{20}$

Step 2: Add the numerators of the fractions. Write the sum over the common denominator.

$16 + 15 = 31$, so 31 is the numerator of your new fraction.
The new fraction is $\frac{31}{20}$.

Step 3: Add the whole numbers.

$1 + 7 = 8$

Step 4: Write the whole number and the fraction beside one another.

$8\frac{31}{20}$

Step 5: If your answer includes an improper fraction, convert the improper fraction to a mixed number and add it to the whole number part of your answer.

You know that $\frac{31}{20}$ is an improper fraction because 31 is greater than 20. Divide 31 by 20 to change the fraction into a mixed number: $\frac{31}{20} = 1\frac{11}{20}$. Then add the mixed number to the whole number 8.

$1\frac{11}{20} + 8 = 9\frac{11}{20}$.

If you aren't sure what the lowest common denominator is, multiply the denominators together. The product of the two denominators will be a common denominator. However, it might not be the lowest common denominator.

Subtracting mixed numbers is very similar to adding them. Follow these steps to subtract mixed numbers.

Step 1: If the fractions do not have a common denominator, change them into fractions with a common denominator.

Step 2: Subtract the numerators of the fractions. Write the difference over the common denominator.

Step 3: Subtract the whole numbers.

Step 4: Write the whole number and the fraction beside one another.

Step 5: If your answer includes an improper fraction, convert the improper fraction to a mixed number and add it to the whole number part of your answer.

Example: $7\frac{3}{4} - 6\frac{1}{2} =$

Step 1: If the fractions do not have a common denominator, change them into fractions with a common denominator.

The fractions $\frac{3}{4}$ and $\frac{1}{2}$ have different denominators, so you need to find a common denominator. You might choose 4 as a common denominator since 2 divides evenly into 4. Then you will have to convert only one of the fractions.

$\frac{1 \times 2}{2 \times 2} = \frac{2}{4}$

Step 2: Subtract the numerators of the fractions. Write the difference over the common denominator.

$3 - 2 = 1$, so 1 is the numerator of your new fraction.
The new fraction is $\frac{1}{4}$.

Step 3: Subtract the whole numbers.

$7 - 6 = 1$

Step 4: Write the whole number and the fraction beside one another.

$1\frac{1}{4}$

Step 5: If your answer includes an improper fraction, convert the improper fraction to a mixed number and add it to the whole number part of your answer. Your answer does not include an improper fraction, so you can skip this step.

In the last example, the first fraction was larger than the second one, so you could simply subtract $\frac{1}{2}$ from $\frac{3}{4}$. But what if the first fraction is smaller than the fraction you are subtracting from it? If the numerator of the first fraction is smaller than the numerator of the second fraction, you will have to regroup. Here's how to regroup with fractions.

Step 1: Borrow 1 from the whole number.
Step 2: Write the 1 as a fraction with the same denominator as the fraction.
Step 3: Add the 1 to the fraction. Your answer will be an improper fraction.
Step 4: Rewrite your subtraction problem. Subtract your mixed numbers as usual.

EXAMPLE: $5\frac{1}{10} - 3\frac{8}{10} =$

You will have to regroup because $\frac{1}{10}$ is smaller than $\frac{8}{10}$. You cannot take 8 away from 1. Follow these steps to regroup the mixed number $5\frac{1}{10}$.

Step 1: Borrow 1 from the whole number.

Reduce the 5 by 1. So your whole number is now 4.

continued from previous page

Step 2: Write the 1 as a fraction with the same denominator as the fraction.

You know that any number over itself is equal to 1. So first look at the denominator of the fraction part of the mixed number. The denominator is 10, so you should write the 1 you have borrowed as $\frac{10}{10}$.

Step 3: Add the 1 to the fraction. Your answer will be an improper fraction.

$$\frac{10}{10} + \frac{1}{10} = \frac{11}{10}$$

Step 4: Write your new whole number and your new fraction side-by-side.

$$4\frac{11}{10}$$

Step 5: Rewrite your subtraction problem. Subtract your mixed numbers as usual.

$$4\frac{11}{10} - 3\frac{8}{10} = 1\frac{3}{10}$$

PRACTICE

Add or subtract the mixed numbers. Reduce your answers to lowest terms. You can check your answers at the end of the section.

26. $7\frac{5}{8} + 3\frac{1}{8} =$

27. $5\frac{1}{3} + 9\frac{2}{3} =$

28. $4\frac{2}{5} + 3\frac{3}{4} =$

29. $6\frac{1}{2} + 9\frac{11}{12} =$

30. $1\frac{1}{6} + 4\frac{3}{4} + 3\frac{1}{2} =$

31. $5\frac{7}{10} - 2\frac{3}{10} =$

32. $20\frac{1}{5} - 6\frac{4}{5} =$

33. $3\frac{1}{2} - 1\frac{8}{9} =$

34. $7\frac{1}{10} - 3\frac{8}{10} =$

35. $9\frac{1}{8} - 5\frac{3}{4} =$

▶ MULTIPLYING AND DIVIDING MIXED NUMBERS

Before you multiply or divide mixed numbers, change them into improper fractions. Then multiply or divide as usual.

Example: $3\frac{2}{3} \times 2\frac{1}{2}$

Step 1: Change the mixed numbers to improper fractions.

$3\frac{2}{3} = \frac{11}{3}$

$2\frac{1}{2} = \frac{5}{2}$

So now you can rewrite the problem as $\frac{11}{3} \times \frac{5}{2} =$

Step 2: Multiply the numerators. Write the product as the numerator of your answer.

$11 \times 5 = 55$

So the top number of your answer is 55.

Step 3: Multiply the denominators. Write the product as the denominator of your answer.

$3 \times 2 = 6$

So your fraction is $\frac{55}{6}$.

Step 4: Change any improper fractions to proper fractions. Reduce your answer to lowest terms.

$$\begin{array}{r} 9 \\ 6\overline{)55} \\ -\underline{54} \\ 1 \end{array}$$

So your answer written in lowest terms is $9\frac{1}{6}$.

Example: $3\frac{2}{3} \div 1\frac{2}{3}$

Step 1: Change the mixed numbers to improper fractions.

$3\frac{2}{3} = \frac{11}{3}$

$1\frac{2}{3} = \frac{5}{3}$

So now you can rewrite the problem as $\frac{11}{3} \div \frac{5}{3} =$

Step 2: Invert the second fraction.

$\frac{5}{3}$ inverts to $\frac{3}{5}$

Step 3: Multiply the fractions.

$\frac{11}{3} \times \frac{3}{5} = \frac{11}{\cancel{3}_1} \times \frac{\cancel{3}^1}{5} = \frac{11}{5}$

Step 4: Change any improper fractions to proper fractions. Reduce your answer to lowest terms.

$$\begin{array}{r} 2 \\ 5\overline{)11} \\ -\underline{10} \\ 1 \end{array}$$

So your answer written in lowest terms is $2\frac{1}{5}$.

PRACTICE

Multiply or divide the mixed numbers. Reduce your answers to lowest terms. You can check your answers at the end of the section.

36. $2\frac{1}{2} \times 3\frac{1}{3} =$

37. $5\frac{1}{3} \times 2\frac{2}{5} =$

38. $4\frac{2}{5} \times 3\frac{3}{4} =$

39. $6\frac{1}{2} \times 9\frac{1}{4} =$

40. $1\frac{1}{6} \times 4\frac{3}{8} =$

41. $5\frac{1}{3} \div 2\frac{3}{5} =$

42. $4\frac{1}{5} \div 1\frac{4}{5} =$

43. $3\frac{1}{2} \div 1\frac{2}{3} =$

44. $5\frac{1}{10} \div 3\frac{1}{2} =$

45. $9\frac{1}{8} \div 5\frac{3}{4} =$

Real World Problems

These problems apply the skills you've learned in Section 2 to every day situations. As you work through these problems, you'll see that the skills you've learned in this section aren't only important for math tests. They are important skills for ordinary questions that come up every day. You can check your answers at the end of the section.

When you divide by a proper fraction, the answer is greater than the number you are dividing. However, when you divide by an improper fraction, the answer is less than the number you are dividing.

1. Toby's mother buys a package of eight granola bars for his lunch each week. If Toby takes one granola bar to school in his lunch each day, what fraction of the granola bars are left over at the end of the school week?
2. Patrick plays the flute. It takes him $3\frac{1}{2}$ minutes to play the new song this week. If Patrick needs to practice the song six times each day, how long will it take him?
3. The directions on an exam allow $2\frac{1}{2}$ hours to answer 50 questions. If you want to spend an equal amount of time on each of the 50 questions, about how much time should you allow for each one?

4. Here is a list of the ingredients needed to make 16 brownies.

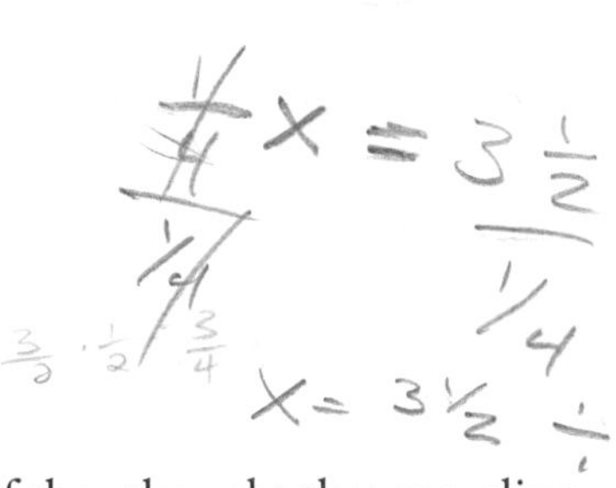

DELUXE BROWNIES

$\frac{2}{3}$ cup butter
5 squares (1 ounce each) unsweetened chocolate
$1\frac{1}{2}$ cups sugar
2 teaspoons vanilla extract
2 eggs
1 cup flour

How much sugar is needed to make 8 brownies?

5. Jackie did a survey for her social studies project. She asked people if they thought that recycling is important. Of the people Jackie talked to, $\frac{7}{8}$ said that recycling is important. Only $\frac{1}{3}$ of the people who said recycling is important said that they also buy things made from recycled materials. What fraction of the people who said recycling is important buys things made from recycled materials?

6. One lap on a particular outdoor track measures a quarter of a mile around. To run a total of three and a half miles, how many complete laps must a person complete?

7. Julien walks three days a week. He keeps track of the number of miles he walks each week. This week he walked $1\frac{1}{2}$ miles on Tuesday, $3\frac{1}{4}$ miles on Thursday, and $2\frac{3}{4}$ miles on Saturday. How many miles did Julien walk this week?

8. Kyril's parents allow him to work 10 hours per week after school and on weekends. This week he has already worked $8\frac{1}{3}$ hours. How many more hours is Kyril allowed to work this week?

9. As part of a school project, Sunita recorded how much gasoline her family put in their car in one month. Her family bought $8\frac{1}{2}$ gallons, $7\frac{3}{4}$ gallons, and $6\frac{7}{10}$ gallons over the month. How much gasoline did Sunita's family put in their car for the month?

10. Jason works in a pizza delivery restaurant. On Friday, the restaurant received $22\frac{1}{2}$ pounds of cheese. Jason used half of the cheese over the weekend. How much cheese is left over for pizzas on Monday?

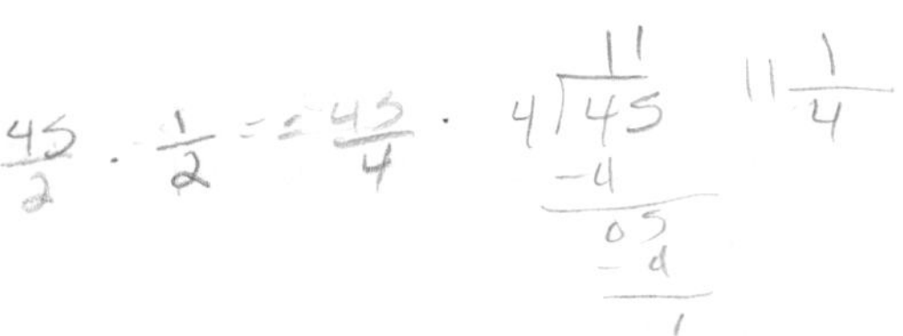

SECTION

2

Answers & Explanations

LESSON 3

1. Three out of the five wedges of the circle are shaded; therefore, the fraction representing the shaded part of the whole is $\frac{3}{5}$.
2. One out of the three wedges of the circle are shaded; therefore, the fraction representing the shaded part of the whole is $\frac{1}{3}$.
3. Six out of the 15 rectangles in the grid are shaded; therefore, the fraction representing the shaded part of the whole is $\frac{6}{15}$.
4. Five out of the nine cylinders are shaded; therefore, the fraction representing the shaded part of the whole is $\frac{5}{9}$.
5. Four out of the five triangles are shaded; therefore, the fraction representing the shaded part of the whole is $\frac{4}{5}$.
6. Three out of the eight rectangles are shaded; therefore, the fraction representing the shaded part of the whole is $\frac{3}{8}$.
7. One of the three circles is shaded; therefore, the fraction representing the shaded part of the whole is $\frac{1}{3}$.
8. First, determine the denominator. There are 20 students in Karen's French class, so the denominator is 20. Three students have visited France, so the fraction of students who have visited France is $\frac{3}{20}$.
9. First, determine the denominator. There are five days in a school week, so the denominator is 5. The cafeteria serves hamburgers one day per week, so the fraction of days each school week that the cafeteria serves hamburgers is $\frac{1}{5}$.

10. First, determine the denominator. The store carries 900 videos, so the denominator is 900. Only 40 of these videos are rated G, so the fraction of the store's videos that is rated G is $\frac{40}{900}$. Reduce the fraction to lowest terms: $\frac{4}{90} \div \frac{2}{2} = \frac{2}{45}$. Thus, $\frac{2}{45}$ of the store's videos are rated G.

11. Twelve will divide evenly into both the numerator and the denominator, so you reduce the fraction by dividing by $\frac{12}{12}$: $\frac{12}{48} \div \frac{12}{12} = \frac{1}{4}$.

12. Seven will divide evenly into both the numerator and the denominator, so you reduce the fraction by dividing by $\frac{7}{7}$: $\frac{7}{49} \div \frac{7}{7} = \frac{1}{7}$.

13. Nine will divide evenly into both the numerator and the denominator, so you reduce the fraction by dividing by $\frac{9}{9}$: $\frac{9}{81} \div \frac{9}{9} = \frac{1}{9}$.

14. Two will divide evenly into both the numerator and the denominator, so you reduce the fraction by dividing by $\frac{2}{2}$: $\frac{42}{88} \div \frac{2}{2} = \frac{21}{44}$.

15. Fifty will divide evenly into both the numerator and the denominator, so you reduce the fraction by dividing by $\frac{50}{50}$: $\frac{150}{250} \div \frac{50}{50} = \frac{3}{5}$.

Since both the numerator and the denominator end in zero, to begin reducing, you can cross out a zero (which is the same as dividing by ten), making the number smaller and easier to manage: $\frac{150}{250}$ becomes $\frac{15}{25}$. Then, you easily see that both numbers are divisible by 5, and you get $\frac{15}{25} \div \frac{5}{5} = \frac{3}{5}$.

16. One-hundred twenty will divide evenly into both the numerator and the denominator, so you reduce the fraction by dividing by $\frac{120}{120}$: $\frac{240}{360} \div \frac{120}{120} = \frac{2}{3}$

Since both the numerator and the denominator end in zero, to begin reducing, you can cross out a zero (which is the same as dividing by ten), making the number smaller and easier to manage: $\frac{240}{360}$ becomes $\frac{24}{36}$. Then, you easily see that both numbers are divisible by 12, and you get $\frac{24}{36} \div \frac{12}{12} = \frac{2}{3}$.

17. Nine will divide evenly into both the numerator and the denominator, so you reduce the fraction by dividing by $\frac{9}{9}$: $\frac{18}{27} \div \frac{9}{9} = \frac{2}{3}$.

18. First, determine the denominator. The total amount of time Jean spent on her project was 10 hours, so the denominator is 10. On Saturday, she spent 2 hours working on the project, so the numerator is 2. The fraction is $\frac{2}{10}$, which when divided by $\frac{2}{2}$ can be reduced to $\frac{1}{5}$. So, Jean worked $\frac{1}{5}$ of the total time it took her to complete the project on Saturday.

19. The total amount of Patrick's savings was \$100, so the denominator is 100. Over the weekend, he spent \$40, so the numerator is 40. The fraction is $\frac{40}{100}$, which when divided by $\frac{20}{20}$ can be reduced to $\frac{2}{5}$. So, Patrick spent $\frac{2}{5}$ of his savings.

20. There was a total of 16 pieces of pizza, so the denominator is 16. The family ate all but 3 slices of the pizza, or 13 (16 pieces total – 3 pieces left) pieces of pizza, so the numerator is 13. Therefore, the fraction of pizza that the family ate is $\frac{13}{16}$.

21. This problem is asking you to raise the fraction $\frac{2}{3}$ to 12ths. You know that you have to multiply 3 (the denominator) by 4 to get 12, so you know that you also need to multiply the numerator by 4. (Remember, to raise a fraction to higher terms, you multiply both the numerator and the denominator by the same number). So, the answer is $\frac{2}{3} \times \frac{4}{4} = \frac{8}{12}$.

22. This problem is asking you to raise the fraction $\frac{5}{8}$ to 32nds. You know that you have to multiply 8 (the denominator) by 4 to get 32, so you know that you also need to multiply the numerator by 4. (Remember, to raise a fraction to higher terms, you multiply both the numerator and the denominator by the same number). So, the answer is $\frac{5}{8} \times \frac{4}{4} = \frac{20}{32}$.

23. This problem is asking you to raise the fraction $\frac{3}{8}$ to 32nds. You know that you have to multiply 8 (the denominator) by 4 to get 32, so you know that you also need to multiply the numerator by 4. (Remember, to raise a fraction to higher terms, you multiply both the numerator and the denominator by the same number). So, the answer is $\frac{3}{8} \times \frac{4}{4} = \frac{12}{32}$.

24. This problem is asking you to raise the fraction $\frac{12}{20}$ to 40ths. You know that you have to multiply 20 (the denominator) by 2 to get 40, so you know that you also need to multiply the numerator by 2. (Remember, to raise a fraction to higher terms, you multiply both the numerator and the denominator by the same number). So, the answer is $\frac{12}{20} \times \frac{2}{2} = \frac{24}{40}$.

25. This problem is asking you to raise the fraction $\frac{9}{10}$ to 50ths. You know that you have to multiply 10 (the denominator) by 5 to get 50, so you know that you also need to multiply the numerator by 5. (Remember, to raise a fraction to higher terms, you multiply both the numerator and the denominator by the same number). So, the answer is $\frac{9}{10} \times \frac{5}{5} = \frac{45}{50}$.

26. This problem is asking you to raise the fraction $\frac{7}{13}$ to 39ths. You know that you have to multiply 13 (the denominator) by 3 to get 39, so you know that you also need to multiply the numerator by 3. (Remember, to raise a fraction to higher terms, you multiply both the numerator and the denominator by the same number). So, the answer is $\frac{7}{13} \times \frac{3}{3} = \frac{21}{39}$.

27. This problem is asking you to raise the fraction $\frac{3}{10}$ to 120ths. You know that you have to multiply 10 (the denominator) by 12 to get 120, so you know that you also need to multiply the numerator by 12. (Remember, to raise a fraction to higher terms, you multiply both the numerator and the denominator by the same number). So, the answer is $\frac{3}{10} \times \frac{12}{12} = \frac{36}{120}$.

28. This problem is asking you to raise the fraction $\frac{20}{50}$ to one-hundredths. You know that you have to multiply 50 (the denominator) by 2 to get 100, so you know that you also need to multiply the numerator by 2. (Remember, to raise a fraction to higher terms, you multiply both the numerator and the denominator by the same number). So, the answer is $\frac{20}{50} \times \frac{2}{2} = \frac{40}{100}$.

29. This problem is asking you to raise the fraction $\frac{2}{9}$ to 810ths. You know that you have to multiply 9 (the denominator) by 90 to get 810, so you know that you also need to multiply the numerator by 90. (Remember, to raise a fraction to higher terms, you multiply both the numerator and the denominator by the same number). So, the answer is $\frac{2}{9} \times \frac{90}{90} = \frac{180}{810}$.

30. This problem is asking you to raise the fraction $\frac{1}{8}$ to 640ths. You know that you have to multiply 8 (the denominator) by 80 to get 640, so you know that you also need to multiply the numerator by 80. (Remember, to raise a fraction to higher terms, you multiply both the numerator and the denominator by the same number). So, the answer is $\frac{1}{8} \times \frac{80}{80} = \frac{80}{640}$.

LESSON 4

1. Since the denominators are the same, this problem is just a simple addition problem. You just add the numerators, 1 + 3 = 4 and put it all over 5. Thus, your answer is $\frac{4}{5}$.

2. Since the denominators are the same, this problem is just a simple addition problem. You just add the numerators, 1 + 1 = 2, and put it all over 3, so your final answer is $\frac{2}{3}$.

3. Since the denominators are the same, this problem is just a simple addition problem. You just add the numerators, 2 + 3 = 5, and put it all over 6, so your final answer is $\frac{5}{6}$.

4. Since the denominators are the same, this problem is just a simple addition problem. You just add the numerators, 7 + 1 + 2 = 10, and put it all over 15, and you get $\frac{10}{15}$. But, you have to reduce your answer to the lowest terms, so you divide by $\frac{5}{5}$. (5 is the largest number that will divide evenly into both 10 and 15.) Thus, your final answer is $\frac{2}{3}$.

5. Since the denominators are the same, this problem is just a simple addition problem. You just add the numerators, 1 + 6 + 3 = 10, and put it all over 20, and you get $\frac{10}{20}$. But, you have to reduce your answer to the lowest terms, so you divide by $\frac{10}{10}$. (10 is the largest number that will divide evenly into both 10 and 20.) Thus, your final answer is $\frac{1}{2}$.

6. Since the denominators are the same, this problem is just a simple subtraction problem. You just subtract the numerators, 8 – 2 = 6, and put it all over 9, and you get $\frac{6}{9}$. But, you have to reduce your answer to the lowest terms, so you divide by $\frac{3}{3}$. (3 is the largest number that will divide evenly into both 6 and 9.) Thus, your final answer is $\frac{2}{3}$.

7. Since the denominators are the same, this problem is just a simple subtraction problem. You just subtract the numerators, 7 – 1 = 6, and put it all over 10, and you get $\frac{6}{10}$. But, you have to reduce your answer to the lowest terms, so you divide by $\frac{2}{2}$. (2 is the largest number that will divide evenly into both 6 and 10.) Thus, your final answer is $\frac{3}{5}$.

8. Since the denominators are the same, this problem is just a simple subtraction problem. You just subtract the numerators, 3 – 1 = 2, and put it all over 11, so your final answer is $\frac{2}{11}$.

9. Since the denominators are the same, this problem is just a simple subtraction problem. You just subtract the numerators, 12 – 2 = 10, and put it all over 25, and you get $\frac{10}{25}$. But, you have to reduce your answer to the lowest terms, so you divide by $\frac{5}{5}$. (5 is the largest number that will divide evenly into both 10 and 25.) Thus, your final answer is $\frac{2}{5}$.

10. Since the denominators are the same, this problem is just a simple subtraction problem. You just subtract the numerators, 4 – 1 = 3, and put it all over 15, and you get $\frac{3}{15}$. But, you have to reduce your answer to the lowest terms, so you divide by $\frac{3}{3}$. (3 is the largest number that will divide evenly into both 3 and 15.) Thus, your final answer is $\frac{1}{5}$.

11. First, because the numbers have different denominators, you have to find a common denominator. Because 5 will divide evenly into 10, you can use 10 as your common denominator and you only have to raise one fraction to higher terms. So, raise $\frac{1}{5}$ to a higher power by multiplying by $\frac{2}{2}$: $\frac{1}{5} \times \frac{2}{2} = \frac{2}{10}$. Now, you just have to add the numerators: 2 + 6 = 8, and put it all over your denominator, 10: $\frac{8}{10}$. Then you just have to reduce. Two is the largest number that goes into both numbers, so you reduce the fraction to your final answer: $\frac{4}{5}$.

12. First, because the numbers have different denominators, you have to find a common denominator. Because 7 will divide evenly into 21, you can use 21 as your common denominator and you only have to raise one fraction to higher terms. So, raise $\frac{2}{7}$ to a higher power by multiplying by $\frac{3}{3}$: $\frac{2}{7} \times \frac{3}{3} = \frac{6}{21}$. Now, you just have to add the numerators: 6 + 3 = 9, and put it all over your denominator, 21: $\frac{9}{21}$. Then you just have to reduce. Three is the largest number that goes into both numbers, so you reduce the fraction to your final answer: $\frac{3}{7}$.

13. First, because the numbers have different denominators, you have to find a common denominator. 60 is the lowest common denominator, so you have to raise both fractions to higher terms. You multiply: $\frac{2}{15} \times \frac{4}{4} = \frac{8}{60}$, and $\frac{5}{20} \times \frac{3}{3} = \frac{15}{60}$. Now, you just have to add the numerators: 8 + 15 = 23. Now, you put 23 over your denominator and get $\frac{23}{60}$. This is your final answer since you can't reduce the fraction any further.

14. First, because the numbers have different denominators, you have to find a common denominator. 60 is the lowest common denominator, so you have to raise both fractions to higher terms. You multiply: $\frac{1}{5} \times \frac{12}{12} = \frac{12}{60}$, and $\frac{6}{12} \times \frac{5}{5} = \frac{30}{60}$. Now, you just have to add the numerators: 12 + 30 = 42. Now, you put 42 over your denominator and get $\frac{42}{60}$. Then you just have to reduce. Six is the largest number that goes into both numbers, so you reduce the fraction to your final answer: $\frac{7}{10}$.

15. First, because the numbers have different denominators, you have to find a common denominator. 12 is the lowest common denominator, so you have to raise all three fractions to higher terms. You multiply: $\frac{1}{2} \times \frac{6}{6} = \frac{6}{12}$, $\frac{1}{4} \times \frac{3}{3} = \frac{3}{12}$, and $\frac{1}{6} \times \frac{2}{2} = \frac{2}{12}$. Now, you just have to add the numerators: 6 + 3 + 2 = 11. Then, you put 11 over your denominator and get $\frac{11}{12}$. This is your final answer since you can't reduce the fraction any further.

16. First, because the numbers have different denominators, you have to find a common denominator. 16 is the lowest common denominator, so you have to raise two of the three fractions to higher terms. You multiply: $\frac{3}{8} \times \frac{2}{2} = \frac{6}{16}$, and $\frac{1}{4} \times \frac{4}{4} = \frac{4}{16}$. Now, you just have to add the numerators: 6 + 4 + 3 = 13. Then, you put 13 over your denominator and get $\frac{13}{16}$. This is your final answer since you can't reduce the fraction any further.

17. First, because the numbers have different denominators, you have to find a common denominator. Because 2 will divide evenly into 4, you can use 4 as your common denominator and you only have to raise one fraction to higher terms. So, raise $\frac{1}{2}$ to a higher power by multiplying by $\frac{2}{2}$: $\frac{1}{2} \times \frac{2}{2} = \frac{2}{4}$. Now, you just have to subtract the numerators: 3 – 2 = 1, and put it all over your denominator, 4: $\frac{1}{4}$. This is your final answer since you can't reduce the fraction any further.

18. First, because the numbers have different denominators, you have to find a common denominator. Because 2 will divide evenly into 6, you can use 6 as your common denominator and you only have to raise one fraction to higher terms. So, raise $\frac{1}{2}$ to a higher power by multiplying by $\frac{3}{3}$: $\frac{1}{2} \times \frac{3}{3} = \frac{3}{6}$. Now, you just have to subtract the numerators: 5 – 3 = 2, and put it all over your denominator, 6: $\frac{2}{6}$. Then you just have to reduce. Three is the largest number that goes into both numbers, so you reduce the fraction to your final answer: $\frac{1}{3}$.

19. First, because the numbers have different denominators, you have to find a common denominator. Because 3 will divide evenly into 9, you can use 9 as your common denominator and you only have to raise one fraction to higher terms. So, raise $\frac{1}{3}$ to a higher power by multiplying by $\frac{3}{3}$: $\frac{1}{3} \times \frac{3}{3} = \frac{3}{9}$. Now, you just have to subtract the numerators: 3 – 1 = 2, and put it all over your denominator, 9: $\frac{2}{9}$. This is your final answer since you can't reduce the fraction any further.

20. First, because the numbers have different denominators, you have to find a common denominator. 20 is the lowest common denominator, so you have to raise both fractions to higher terms. You multiply: $\frac{1}{4} \times \frac{5}{5} = \frac{5}{20}$, and $\frac{1}{5} \times \frac{4}{4} = \frac{4}{20}$. Now, you just have to subtract the numerators: 5 – 4 = 1. Then, you put 1 over your denominator and get $\frac{1}{20}$. This is your final answer since you can't reduce the fraction any further.

21. First, because the numbers have different denominators, you have to find a common denominator. 18 is the lowest common denominator, so you have to raise both fractions to higher terms. You multiply: $\frac{1}{2} \times \frac{9}{9} = \frac{9}{18}$, and $\frac{1}{9} \times \frac{2}{2} = \frac{2}{18}$. Now, you just have to subtract the numerators: 9 – 2 = 7. Then, you put 7 over your denominator and get $\frac{7}{18}$. This is your final answer since you can't reduce the fraction any further.

22. First, because the numbers have different denominators, you have to find a common denominator. 4 will divide evenly into 20, so you just have to raise one fraction to higher terms. You multiply: $\frac{3}{4} \times \frac{5}{5} = \frac{15}{20}$. Then, you just have to subtract the numerators: 15 – 6 = 9. Finally, you put 9 over your denominator and get $\frac{9}{20}$. This is your final answer since you can't reduce the fraction any further.

23. First, because the numbers have different denominators, you have to find a common denominator. 72 is the lowest common denominator, so you have to raise both fractions to higher terms. You multiply: $\frac{7}{8} \times \frac{9}{9} = \frac{63}{72}$, and $\frac{3}{9} \times \frac{8}{8} = \frac{24}{72}$. Now, you just have to subtract the numerators: 63 – 24 = 39. Now, you put 39 over your denominator and get $\frac{39}{72}$. Then you just have to reduce. Three is the largest number that goes into both numbers, so you reduce the fraction to your final answer: $\frac{13}{24}$.

24. First, because the numbers have different denominators, you have to find a common denominator. 4 will divide evenly into 16, so you just have to raise one fraction to higher terms. You multiply: $\frac{1}{4} \times \frac{4}{4} = \frac{4}{16}$. Then, you just have to subtract the numerators: 7 – 4 = 3. Finally, you put

3 over your denominator and get $\frac{3}{16}$. This is your final answer since you can't reduce the fraction any further.

25. First, because the numbers have different denominators, you have to find a common denominator. 30 is the lowest common denominator, so you have to raise both fractions to higher terms. You multiply: $\frac{9}{10} \times \frac{3}{3} = \frac{27}{30}$, and $\frac{12}{15} \times \frac{2}{2} = \frac{24}{30}$. Now, you just have to subtract the numerators: 27 – 24 = 3. Now, you put 3 over your denominator and get $\frac{3}{30}$. Then you just have to reduce. Three is the largest number that goes into both numbers, so you reduce the fraction to your final answer: $\frac{1}{10}$.

LESSON 5

1. You just have to multiply across the top and bottom of the fraction: $\frac{1}{3} \times \frac{1}{2} = \frac{1}{6}$.
2. You just have to multiply across the top and bottom of the fraction: $\frac{2}{5} \times \frac{1}{5} = \frac{2}{25}$.
3. You just have to multiply across the top and bottom of the fraction: $\frac{3}{12} \times \frac{4}{20} = \frac{12}{240}$. Then you have to reduce. Twelve goes evenly into both numbers, so you divide the top and bottom by twelve and get your final answer: $\frac{1}{20}$.
4. You just have to multiply across the top and bottom of the fraction: $\frac{6}{7} \times \frac{3}{4} = \frac{18}{28}$. Then you have to reduce. Two goes evenly into both numbers, so you divide the top and bottom by 2 and get your final answer: $\frac{9}{14}$.
5. You just have to multiply across the top and bottom of the fraction: $\frac{2}{20} \times \frac{6}{8} = \frac{12}{240}$. Then you have to reduce. Twelve goes evenly into both numbers, so you divide the top and bottom by 12 and get your final answer: $\frac{1}{20}$.
6. You just have to multiply across the top and bottom of the fraction: $\frac{7}{21} \times \frac{3}{7} = \frac{21}{147}$. Then you have to reduce. Twenty-one goes evenly into both numbers, so you divide the top and bottom by 12 and get your final answer: $\frac{1}{7}$.
7. You just have to multiply across the top and bottom of the fraction: $\frac{3}{100} \times \frac{10}{17} = \frac{30}{1700}$. Then you have to reduce. Ten goes evenly into both numbers, so you divide the top and bottom by ten and get your final answer: $\frac{3}{170}$.

You can save a step in reducing if you cancel some zeros when you first set up the problem:

$\frac{3}{100} \times \frac{10}{17}$. **You have simplified your problem to** $\frac{3}{10} \times \frac{1}{17} = \frac{3}{170}$.

8. You just have to multiply across the top and bottom of the fraction: $\frac{3}{7} \times \frac{7}{9} = \frac{21}{63}$. Then you have to reduce. 21 goes evenly into both numbers, so you divide the top and bottom by 21, and get your final answer: $\frac{1}{3}$.

You can save a step in reducing if you cancel when you first set up the problem: $\frac{3}{7} \times \frac{7}{9}$. You can cancel the three and the nine and the sevens:

$$\frac{\overset{1}{\cancel{3}}}{\underset{1}{\cancel{7}}} \times \frac{\overset{1}{\cancel{7}}}{\underset{3}{\cancel{9}}} = \frac{1}{3}, \text{ your final answer.}$$

9. The problem is asking you to multiply $\frac{5}{8}$ and $\frac{6}{15}$. First, multiply the numerators: $5 \times 6 = 30$. So, the top number of your answer is 30. Then, multiply the denominators: $8 \times 15 = 120$. The bottom number is 120, so your fraction is $\frac{30}{120}$. Reduce the fraction to lowest terms. The numbers 30 and 120 can both be divided by 30. So, your answer written in lowest terms is $\frac{1}{4}$.

10. The problem is asking you to multiply $\frac{8}{9}$ and $\frac{5}{6}$. First, multiply the numerators: $8 \times 5 = 40$. So the top number of your answer is 40. Then, multiply the denominators: $9 \times 6 = 54$. The bottom number is 54, so your fraction is $\frac{40}{54}$. Reduce the fraction to lowest terms. The numbers 40 and 54 can both be divided by 2. So your answer written in lowest terms is $\frac{20}{27}$.

11. She needs $\frac{1}{2}$ of $\frac{3}{4}$ cup of sugar, so you need to multiply $\frac{1}{2} \times \frac{3}{4}$. First, multiply the numerators: $1 \times 3 = 3$. Then, multiply the denominators: $2 \times 4 = 8$. Thus, Elizabeth needs $\frac{3}{8}$ cup of sugar.

12. Jake ran $\frac{1}{2}$ of $\frac{1}{4}$ mile. First, multiply the numerators: $1 \times 1 = 1$. Then, multiply the denominators: $2 \times 4 = 8$. Thus, Jake ran $\frac{1}{8}$ mile.

13. Invert the second fraction and multiply the two fractions: $\frac{1}{4} \times \frac{3}{1} = \frac{3}{4}$.

14. Invert the second fraction and multiply the two fractions: $\frac{1}{3} \times \frac{2}{1} = \frac{2}{3}$.

15. Invert the second fraction and multiply the two fractions: $\frac{1}{2} \times \frac{6}{5} = \frac{6}{10}$. Then, write the answer in lowest terms. Both 6 and 10 can be divided by 2, so the final answer is $\frac{3}{5}$.

16. Invert the second fraction and multiply the two fractions: $\frac{3}{9} \times \frac{10}{8} = \frac{30}{72}$. Then, write the answer in lowest terms. Both 30 and 72 can be divided by 6, so the final answer is $\frac{5}{12}$.

17. Invert the second fraction and multiply the two fractions: $\frac{1}{8} \times \frac{3}{1} = \frac{3}{8}$.

18. Invert the second fraction and multiply the two fractions: $\frac{1}{9} \times \frac{6}{1} = \frac{6}{9}$. Then, write the answer in lowest terms. Both 6 and 9 can be divided by 3, so the final answer is $\frac{2}{3}$.

19. Invert the second fraction and multiply the two fractions: $\frac{2}{7} \times \frac{5}{2}$. There is a 2 in the numerator of the first fraction and in the denominator of the second fraction, so you can cancel by dividing by 2:

$\frac{\overset{1}{\cancel{2}}}{7} \times \frac{5}{\underset{1}{\cancel{2}}}$

You have simplified your problem to $\frac{1}{7} \times \frac{5}{1} = \frac{5}{7}$. The final answer is $\frac{5}{7}$.

20. Invert the second fraction and multiply the two fractions: $\frac{1}{8} \times \frac{4}{3}$. Cancel by dividing by 4:

$\frac{1}{\underset{2}{\cancel{8}}} \times \frac{\overset{1}{\cancel{4}}}{3} = \frac{1}{6}$

The final answer is $\frac{1}{6}$.

21. Invert the second fraction and multiply the two fractions: $\frac{2}{25} \times \frac{5}{4}$. Cancel by dividing by 5 and 2:

$\frac{\overset{1}{\cancel{2}}}{\underset{5}{\cancel{25}}} \times \frac{\overset{1}{\cancel{5}}}{\underset{2}{\cancel{4}}} = \frac{1}{10}$

The final answer is $\frac{1}{10}$.

22. Invert the second fraction and multiply the two fractions: 1\20 × 10\9. Cancel as follows:

$\frac{1}{\underset{2}{\cancel{20}}} \times \frac{\overset{1}{\cancel{10}}}{9} = \frac{1}{18}$

The final answer is $\frac{1}{18}$.

23. Invert the second fraction and multiply the two fractions: $\frac{1}{60} \times \frac{12}{5}$. Cancel as follows:

$$\frac{1}{\cancel{60}_5} \times \frac{\cancel{12}^1}{5} = \frac{1}{25}$$

The final answer is $\frac{1}{25}$.

24. First, set up the problem: $\frac{5}{1} \div \frac{1}{3}$. Then, invert the second fraction and multiply the two fractions: $\frac{5}{1} \times \frac{3}{1} = 15$. Thus, Jason can make 15 burgers using the meat.

25. First, set up the problem: $\frac{10}{1} \div \frac{2}{3}$. Then, invert the second fraction and multiply the two fractions: $\frac{10}{1} \times \frac{3}{2}$. Cancel as follows:

$$\frac{\cancel{10}^5}{1} \times \frac{3}{\cancel{2}_1} = \frac{15}{1} = 15$$

Thus, Thuy can make 15 batches of cookies.

LESSON 6

1. $\frac{5}{2}$ or $2\frac{1}{2}$
2. $\frac{9}{8}$ or $1\frac{1}{8}$
3. $\frac{8}{3}$ or $2\frac{2}{3}$
4. $\frac{10}{6}$ or $1\frac{2}{3}$
5. $\frac{7}{2}$ or $3\frac{1}{2}$
6. Divide 7 (the numerator) by 7 (the denominator) to get the whole number: 1.
7. Divide 15 by 3 to get the whole number: 5.
8. Divide 5 by 4 to get the mixed number: $1\frac{1}{4}$.
9. Divide 36 by 6 to get the whole number: 6.
10. Divide 7 by 5 to get the mixed number: $1\frac{2}{5}$.
11. Divide 29 by 4 to get the mixed number: $7\frac{1}{4}$.
12. Divide 50 by 7 to get the mixed number: $7\frac{1}{7}$.
13. Divide 500 by 50 to get the whole number: 10.
14. Divide 14 by 4 to get the mixed number: $3\frac{1}{2}$.
15. Divide 65 by 60 to get the mixed number: $1\frac{1}{12}$.

16. To convert a whole number to an improper fraction, you write the number over 1: $\frac{17}{1}$.

17. Multiply the whole number part of the mixed number by the denominator of the fraction (6 × 2), and then add the numerator of the fraction to the product ((6 × 2) + 1 = 13). Next, write your answer as the numerator and put it all over the original denominator. So, you get $\frac{13}{2}$.

18. Multiply the whole number part of the mixed number by the denominator of the fraction (3 × 4), and then add the numerator of the fraction to the product ((3 × 4) + 3). Then, write your answer as the numerator and put it all over the original denominator. So, you get $\frac{15}{4}$.

19. Multiply the whole number part of the mixed number by the denominator of the fraction (5 × 6), and then add the numerator of the fraction to the product ((5 × 6) + 1). Then, write your answer as the numerator and put it all over the original denominator. So, you get $\frac{31}{6}$.

20. Multiply the whole number part of the mixed number by the denominator of the fraction (7 × 5), and then add the numerator of the fraction to the product ((7 × 5) + 4). Then, write your answer as the numerator and put it all over the original denominator. So, you get $\frac{39}{5}$.

21. Multiply the whole number part of the mixed number by the denominator of the fraction (12 × 3), and then add the numerator of the fraction to the product ((12 × 3) + 2). Then, write your answer as the numerator and put it all over the original denominator. So, you get $\frac{38}{3}$.

22. Multiply the whole number part of the mixed number by the denominator of the fraction (8 × 7), and then add the numerator of the fraction to the product ((8 × 7) + 6). Then, write your answer as the numerator and put it all over the original denominator. So, you get $\frac{62}{7}$.

23. Multiply the whole number part of the mixed number by the denominator of the fraction (1 × 50), and then add the numerator of the fraction to the product ((1 × 50) + 3). Then, write your answer as the numerator and put it all over the original denominator. So, you get $\frac{53}{50}$.

24. Multiply the whole number part of the mixed number by the denominator of the fraction (2 × 40), then add the numerator of the fraction to the product ((2 × 40) + 23). Next, write your answer as the numerator and put it all over the original denominator. So, you get $\frac{103}{40}$.

25. Multiply the whole number part of the mixed number by the denominator of the fraction (6 × 10), then add the numerator of the fraction to the product ((6 × 10) + 4). Next, write your answer as the numerator and put it all over the original denominator. So, you get $\frac{64}{10}$.

26. The fractions have the same denominator, so you just add the numerators of the fractions: 5 + 1 = 6. Then, write this number over the denominator: $\frac{6}{8}$. Next, add the whole numbers: 7 + 3 = 10. Add the fraction to the whole number: $10\frac{6}{8}$. Finally, reduce the fraction to lowest terms. Your final answer is $10\frac{3}{4}$.

27. The fractions have the same denominator, so you just add the numerators of the fractions: 1 + 2 = 3. Then, write this number over the denominator: $\frac{3}{3}$. You know that $\frac{3}{3} = 1$, so add the whole numbers: 1 + 5 + 9 to get the final answer, 15.

28. First, convert the fractions to fractions with a common denominator: $\frac{2}{5}$ becomes $\frac{8}{20}$, and $\frac{3}{4}$ becomes $\frac{15}{20}$. Then, add the fractions together $\frac{8}{20} + \frac{15}{20} = \frac{23}{20}$. This improper fraction becomes the mixed number $1\frac{3}{20}$. Next, add the mixed number and the whole numbers: $1\frac{3}{20} + 4 + 3 = 8\frac{3}{20}$. Your final answer is $8\frac{3}{20}$.

29. First, convert the fractions to fractions with a common denominator: $\frac{1}{2}$ becomes $\frac{6}{12}$, and $\frac{11}{12}$ stays as it is. Then, add the fractions together: $\frac{6}{12} + \frac{11}{12} = \frac{17}{12}$. This improper fraction becomes the mixed number $1\frac{5}{12}$. Next, add the mixed number and the whole numbers: $1\frac{5}{12} + 6 + 9 = 16\frac{5}{12}$. Your final answer is $16\frac{5}{12}$.

30. First, convert the fractions to fractions with a common denominator: $\frac{1}{6}$ becomes $\frac{2}{12}$, $\frac{3}{4}$ becomes $\frac{9}{12}$, and $\frac{1}{2}$ becomes $\frac{6}{12}$. Then, add the fractions together: $\frac{2}{12} + \frac{9}{12} + \frac{6}{12} = \frac{17}{12}$. This improper fraction becomes the mixed number $1\frac{5}{12}$. Next, add the mixed number and the whole numbers: $1\frac{5}{12} + 1 + 4 + 3 = 9\frac{5}{12}$. Your final answer is $9\frac{5}{12}$.

31. The fractions have the same denominator, so you just subtract the numerators of the fractions: $7 - 3 = 4$. Then, write this number over the denominator: $\frac{4}{10}$. Then, subtract the whole numbers: $5 - 2 = 3$. Next, recombine the fraction and the whole number: $3\frac{4}{10}$. Reduce the fraction to lowest terms: $\frac{4}{10}$ becomes $\frac{2}{5}$. Your final answer is $3\frac{2}{5}$.

32. Since $\frac{1}{5}$ is smaller than $\frac{4}{5}$, you begin by regrouping. Borrow 1 from the 20 and rewrite the fraction as $\frac{6}{5}$. Then, subtract the fractions: $\frac{6}{5} - \frac{4}{5} = \frac{2}{5}$. Next, subtract the whole numbers. Remember you borrowed 1 from the 20, so it's now 19: $19 - 6 = 13$. Write your new whole number and your new fraction side-by-side: $13\frac{2}{5}$. Your final answer is $13\frac{2}{5}$.

33. First, convert the fractions to fractions with a common denominator: $\frac{1}{2}$ becomes $\frac{9}{18}$, and $\frac{8}{9}$ becomes $\frac{16}{18}$. Since $\frac{9}{18}$ is smaller than $\frac{16}{18}$, you have to regroup. Borrow 1 from the 3 and rewrite the first fraction as $\frac{27}{18}$. Now you can subtract the fractions: $\frac{27}{18} - \frac{16}{18} = \frac{11}{18}$. Next, subtract the whole numbers. Remember you borrowed 1 from the 3, so it's now 2: $2 - 1 = 1$. Write your new whole number and your new fraction side-by-side: $1\frac{11}{18}$. Your final answer is $1\frac{11}{18}$.

34. Since $\frac{1}{10}$ is smaller than $\frac{8}{10}$, begin by regrouping. Borrow 1 from the 7 and rewrite the fraction as $\frac{11}{10}$. Then, subtract the fractions: $\frac{11}{10} - \frac{8}{10} = \frac{3}{10}$. Next, subtract the whole numbers. Remember you borrowed 1 from the 7: $6 - 3 = 3$. Write your new whole number and your new fraction side-by-side: $3\frac{3}{10}$. Your final answer is $3\frac{3}{10}$.

35. First, convert the fractions to fractions with a common denominator: $\frac{3}{4}$ becomes $\frac{6}{8}$. Since $\frac{1}{8}$ is smaller than $\frac{6}{8}$, begin by regrouping. Borrow 1 from the 9 and rewrite the fraction as $\frac{9}{8}$. Then, subtract the fractions: $\frac{9}{8} - \frac{6}{8} = \frac{3}{8}$. Next, subtract the whole numbers. Remember you borrowed 1 from the 9, so $8 - 5 = 3$. Write your new whole number and your new fraction side-by-side: $3\frac{3}{8}$. Your final answer is $3\frac{3}{8}$.

36. First, change the mixed numbers to improper fractions: $2\frac{1}{2}$ becomes $\frac{5}{2}$, and $3\frac{1}{3}$ becomes $\frac{10}{3}$. Then, you can multiply the fractions more easily: $\frac{5}{2} \times \frac{10}{3} = \frac{50}{6}$. Next, you change this improper fraction to a mixed number: $\frac{50}{6} = 8\frac{2}{6}$. Finally, you can reduce the fraction to the lowest terms, so your final answer is $8\frac{1}{3}$.

37. First, change the mixed numbers to improper fractions: $5\frac{1}{3}$ becomes $\frac{16}{3}$, and $2\frac{2}{5}$ becomes $\frac{12}{5}$. Then, you can multiply the fractions more easily: $\frac{16}{3} \times \frac{12}{5} = \frac{192}{15}$. Next, you change the improper fraction to a mixed number: $\frac{192}{15} = 12\frac{12}{15}$. Finally, you can reduce the fraction to lowest terms: $\frac{12}{15} = \frac{4}{5}$, so your final answer is $12\frac{4}{5}$.

38. First, change the mixed numbers to improper fractions: $4\frac{2}{5}$ becomes $\frac{22}{5}$, and $3\frac{3}{4}$ becomes $\frac{15}{4}$. Then, you can multiply the fractions more easily: $\frac{22}{5} \times \frac{15}{4}$. Use canceling to simplify the multiplication further:

$$\frac{\overset{11}{\cancel{22}}}{\underset{1}{\cancel{5}}} \times \frac{\overset{3}{\cancel{15}}}{\underset{2}{\cancel{4}}} = \frac{33}{2}$$

Next, you change the improper fraction to a proper fraction: $\frac{33}{2} = 16\frac{1}{2}$. Your final answer is $16\frac{1}{2}$.

39. First, change the mixed numbers to improper fractions: $6\frac{1}{2}$ becomes $\frac{13}{2}$, and $9\frac{1}{4}$ becomes $\frac{37}{4}$. Then, you can multiply the fractions more easily: $\frac{13}{2} \times \frac{37}{4} = \frac{481}{8}$. Next, you change the improper fraction to a mixed number: $\frac{481}{8} = 60\frac{1}{8}$. Your final answer is $60\frac{1}{8}$.

40. First, change the mixed numbers to improper fractions: $1\frac{1}{6}$ becomes $\frac{7}{6}$, and $4\frac{3}{8}$ becomes $\frac{35}{8}$. Then you can multiply the fractions more easily: $\frac{7}{6} \times \frac{35}{8} = \frac{245}{48}$. Next you change the improper fraction to a proper fraction: $\frac{245}{48} = 5\frac{5}{48}$. The final answer is $5\frac{5}{48}$.

41. First, change the mixed numbers to improper fractions: $5\frac{1}{3}$ becomes $\frac{16}{3}$, and $2\frac{3}{5}$ becomes $\frac{13}{5}$. Then, you invert the second fraction and multiply: $\frac{16}{3} \times \frac{5}{13} = \frac{80}{39}$. Next, you change the improper fraction to a proper fraction: $\frac{80}{39} = 2\frac{2}{39}$. Your final answer is $2\frac{2}{39}$.

42. First, change the mixed numbers to improper fractions: $4\frac{1}{5}$ becomes $\frac{21}{5}$, and $1\frac{4}{5}$ becomes $\frac{9}{5}$. Then, you invert the second fraction and multiply. Use canceling to simplify the multiplication.

$$\frac{\overset{7}{\cancel{21}}}{\underset{1}{\cancel{5}}} \times \frac{\overset{1}{\cancel{5}}}{\underset{3}{\cancel{9}}} = \frac{7}{3}$$

Next, you change the improper fraction to a proper fraction: $\frac{7}{3} = 2\frac{1}{3}$. Your final answer is $2\frac{1}{3}$.

43. First, change the mixed numbers to improper fractions: $3\frac{1}{2}$ becomes $\frac{7}{2}$, and $1\frac{2}{3}$ becomes $\frac{5}{3}$. Then, you invert the second fraction and multiply: $\frac{7}{2} \times \frac{3}{5} = \frac{21}{10}$. Next, you change the improper fraction to a proper fraction: $\frac{21}{10} = 2\frac{1}{10}$. Your final answer is $2\frac{1}{10}$.

44. First, change the mixed numbers to improper fractions: $5\frac{1}{10}$ becomes $\frac{51}{10}$, and $3\frac{1}{2}$ becomes $\frac{7}{2}$. Then, you invert the second fraction and multiply. Use canceling to simplify the multiplication.

$$\frac{51}{\not{10}_{5}} \times \frac{\not{2}^{1}}{7} = \frac{51}{35}$$

Next, you change the improper fraction to a proper fraction: $\frac{51}{35} = 1\frac{16}{35}$. Your final answer is $1\frac{16}{35}$.

45. First, change the mixed numbers to improper fractions: $9\frac{1}{8}$ becomes $\frac{73}{8}$, and $5\frac{3}{4}$ becomes $\frac{23}{4}$. Then, you invert the second fraction and multiply. Use canceling to simplify the multiplication.

$$\frac{73}{\not{8}_{2}} \times \frac{\not{4}^{1}}{23} = \frac{73}{46}$$

Next, you change the improper fraction to a proper fraction: $\frac{73}{46} = 1\frac{27}{46}$. Your final answer is $1\frac{27}{46}$.

REAL WORLD PROBLEMS

1. Toby takes $\frac{5}{8}$ of the package of granola bars in his school lunch each week, so he has $\frac{3}{8}$ $(\frac{8}{8} - \frac{5}{8})$ left over.

2. First, set the problem up as follows: $\frac{6}{1} \times 3\frac{1}{2}$. Change the mixed number to an improper fraction and cancel as shown below.

$$\frac{\not{6}^{3}}{1} \times \frac{7}{\not{2}_{1}} = \frac{21}{1} = 21$$

Thus, it takes Patrick 21 minutes to practice the song six times.

3. First, set the problem up as follows: $\frac{5}{2} \div 50$. Change the mixed number to an improper fraction and invert it. Then, cancel as shown below.

$$\frac{\overset{1}{\cancel{5}}}{2} \times \frac{1}{\underset{10}{\cancel{50}}} = \frac{1}{20}$$

So, you should spend $\frac{1}{20}$ of an hour on each question. To calculate, the number of minutes you should spend on each question: $\frac{1}{20} \times \frac{60}{1}$ (there are 60 minutes in one hour) $= \frac{60}{20}$. Write $\frac{60}{20}$ as a proper fraction to get 3. Thus, you should spend three minutes on each problem.

A shortcut for this problem is to divide the total number of minutes ($2\frac{1}{2} \times 60 = 150$) by the number of questions (50): $150 \div 50 = 3$. Thus, you should spend three minutes on each problem.

4. First, set the problem up as follows: $1\frac{1}{2} \div 2$. Change the mixed number to an improper fraction. Change the whole number to a fraction and invert it. Then, mulitply: $\frac{3}{2} \times \frac{1}{2} = \frac{3}{4}$. Thus, $\frac{3}{4}$ cup of sugar is needed to make eight brownies.

5. Set the problem up as follows and multiply: $\frac{7}{8} \times \frac{1}{3} = \frac{7}{24}$. Thus, $\frac{7}{24}$ of the people surveyed think recycling is important *and* buy recycled products.

6. First, set the problem up as follows: $3\frac{1}{2} \div \frac{1}{4}$. Change the mixed number to an improper fraction and invert the second fraction. Then, cancel as shown below.

$$\frac{7}{\underset{1}{\cancel{2}}} \times \frac{\overset{2}{\cancel{4}}}{1} = \frac{14}{1} = 14$$

It takes 14 complete laps to make $3\frac{1}{2}$ miles.

7. First, set the problem up as follows: $1\frac{2}{4} + 3\frac{1}{4} + 2\frac{3}{4} = 6 + \frac{6}{4}$. Convert $\frac{6}{4}$ to a mixed number: $\frac{6}{4} = 1\frac{2}{4}$. Add the whole numbers and reduce the fraction to lowest terms:

$6 + 1 = 7$
$\frac{2}{4} = \frac{1}{2}$
$7 + \frac{1}{2} = 7\frac{1}{2}$
He walked $7\frac{1}{2}$ miles last week.

8. First, set the problem up as follows: $10 - 8\frac{1}{3}$. You will have to regroup. Borrow 1 from the 10 and write the 1 as a fraction with the same denominator as the fraction: $9\frac{3}{3} - 8\frac{1}{3}$. Then, subtract: $9\frac{3}{3} - 8\frac{1}{3} = 1\frac{2}{3}$ hours. Kyril can work $1\frac{2}{3}$ hours more this week.

9. First, set the problem up as follows and add: $8\frac{10}{20} + 7\frac{15}{20} + 6\frac{14}{20} = 21 + \frac{39}{20}$. Convert the improper fraction to a mixed number: $\frac{39}{20} = 1\frac{19}{20}$. Add the whole number and the mixed number: $21 + 1\frac{19}{20} = 22\frac{19}{20}$. Thus, they put $22\frac{19}{20}$ gallons in their car over the month.

10. First, set the problem up as follows: $22\frac{1}{2} \times \frac{1}{2}$. Change the mixed number to an improper fraction and multiply: $\frac{45}{2} \times \frac{1}{2} = \frac{45}{4}$. Convert the improper fraction to a mixed number. There are $11\frac{1}{4}$ pounds of cheese left over for Monday.

SECTION

3

Decimals

THIS SECTION WILL introduce you to the idea of decimals. You'll come to understand that a decimal is a special kind of fraction that you use every day when you deal with measurements or money. You'll sharpen your skills so you know them cold! You will learn what decimals are and how to read them. You'll compare them, convert them to fractions, and you will also learn how to perform mathematical operations with them.

Adding and Subtracting Decimals

LESSON SUMMARY

In Section II, you learned how to work with parts of a whole that are represented by fractions. In this lesson, you will begin working with decimals—another way of writing parts of a whole. You will learn what decimals are and how to read them. You will also learn how to write decimals so they are easy to add and subtract.

If you've ever gone shopping, then you are familiar with decimals. We use decimals to represent amounts of money. Like fractions, decimals represent parts of whole numbers. For example, you know that $1.50 is neither one whole dollar nor two whole dollars. It's one dollar and one-half of another dollar. Another way to write 1.50 is $1\frac{1}{2}$.

HOW TO READ A DECIMAL

Notice that *decimals* are numbers written with a dot, or a period, either to the far left or somewhere in the middle. The dot is called a *decimal point.* The numbers to the left of the decimal point are whole numbers. Those to the right of the decimal point are fractions, or parts, of whole numbers.

Although you can always convert a fraction to a decimal by dividing the numerator by the denominator, it's a good idea to know common decimal and fraction equivalents for standardized tests. Here are some common decimals and fractions you might want to learn.

DECIMAL AND FRACTION EQUIVALENTS TO KNOW

Fraction	Decimal
$\frac{1}{100}$	0.01
$\frac{1}{10}$	0.1
$\frac{1}{5}$	0.2
$\frac{1}{4}$	0.25
$\frac{1}{3}$	0.33
$\frac{1}{2}$	0.5
$\frac{2}{3}$	0.67
$\frac{3}{4}$	0.75
$\frac{4}{5}$	0.80
$\frac{9}{10}$	0.90

You probably already know that each digit in the number 1,234 represents a place value. A *place value* is a position in the number. So, for example, the 1 in 1,234 stands for 1 thousand. The 2 stands for 2 hundreds. The 3 stands for 3 tens. And the 4 stands for 4 ones. These are the place values that occur to the *left* of a decimal point. Each digit to the *right* of a decimal point also has a place value. The names and positions of several place values are shown below.

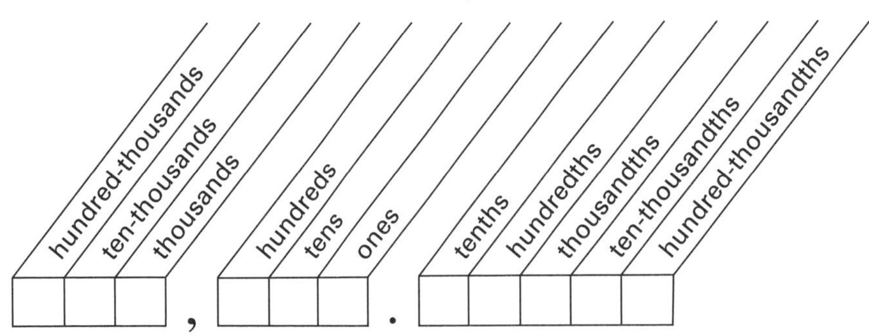

When you see a decimal, here's how to read it.

Step 1: Begin reading from left to right. Read the part of the number that is to the left of the decimal point as you would any other whole number.

Step 2: Read the decimal point as the word *and*.

Step 3: Read the number to the right of the decimal point as you would any other number. But then follow it with the name of the decimal. You can determine the name of the decimal by counting the number of digits to the right of the decimal point.

Example: Write out the following decimal in words: 12.304

Step 1: Begin reading from left to right. Read the part of the number that is to the left of the decimal point as you would any other whole number. The number to the left of the decimal is 12. So you would write (or say if you were reading aloud): "twelve."

Step 2: Read the decimal point as the word *and*. So you would next write "and."

Step 3: Read the number to the right of the decimal point as you would any other number. But then follow it with the name of the decimal. There are three numbers to the right of the decimal point, so the place value is called thousandths. You would write: "three hundred four thousandths."

So the decimal 12.304 can be written in the following words: Twelve and three hundred four thousandths.

Notice that the decimal in the last example has a zero in the middle: 12.304. This zero happens to be in the hundredths place. It tells you that there are no hundredths in the number. However, there are tenths and thousandths, so the zero serves as a *placeholder* between the 3 and the 4. When a zero falls in between two numbers, it serves as a placeholder, and it affects the value of the number.

You can also add zeros at the beginning and end of a number, but they do not affect the value of the number. For example, the following numbers are all equal:

0012.304
012.304
12.3040
12.30400
12.304000

Example: Write the following number as a decimal: forty-three and sixty-seven hundredths.

To write this number as a decimal, you just have to work backwards. It's helpful to first locate the decimal point by looking for the word *and.* Then, write the whole number on each side of the decimal: 43.67. Check your work by making sure that your digits fall in the correct positions.

What if the number in the last example had been: forty-three and sixty-seven thousandths? You might have begun by writing 43.67. But when you counted the digits to the right of the decimal point, you would have realized that you had written sixty-seven hundredths instead of sixty-seven thousandths. You need to add a zero after the decimal point: 43.067. That's why it's important to check your work by counting the digits and making sure that the correct place values are represented in your answer.

PRACTICE

Write out the following decimals in words. You can check your answers at the end of the section.

1. 0.05
2. 0.2
3. 0.22
4. 1.234
5. 25.89
6. 13.90
7. 100.001
8. 45.102
9. 7.8035
10. 2.030

Write the following numbers as decimals. You can check your answers at the end of the section.

11. thirteen hundredths

12. four tenths

13. twenty-five thousandths

14. six hundredths

15. six thousandths

16. three and twenty-eight thousandths

17. eighteen and twelve hundredths

18. twelve ten thousandths

19. twenty-five and six hundred five thousandths

20. sixty-five and one hundred five ten thousandths

ADDING DECIMALS

Follow these steps when adding decimals.

Step 1: Write the numbers so that the decimal points are lined up.
Step 2: Make sure that all the decimals have the same number of digits to the right of the decimal point by adding zeros to the end of shorter decimals.
Step 3: Write the decimal point in the answer so that it lines up with the decimal points in the problem.
Step 4: Add the decimals just as you would if you were adding whole numbers.

The position of the decimal point in a number makes a big difference in its value. Always line up the decimal points before adding or subtracting decimals. This will help you put the decimal point in the correct place and get the correct answer.

Example: 9.23 + 6.02 + 1.1

Step 1: Write the numbers so that the decimal points are lined up.

$$\begin{array}{r} 9.23 \\ 6.02 \\ \underline{+1.1} \end{array}$$

Step 2: Make sure that all the decimals have the same number of digits to the right of the decimal point by adding zeros to the end of shorter decimals.

$$\begin{array}{r} 9.23 \\ 6.02 \\ \underline{+1.10} \end{array}$$

Step 3: Write the decimal point in the answer so that it lines up with the decimal points in the problem.

$$\begin{array}{r} 9.23 \\ 6.02 \\ \underline{+1.10} \\ . \end{array}$$

Step 4: Add the decimals just as you would if you were adding whole numbers.

$$\begin{array}{r} 9.23 \\ 6.02 \\ \underline{+1.10} \\ 16.35 \end{array}$$

The correct answer is 16.35.

How would you add a whole number and a decimal together? You would follow the same steps.

EXAMPLE: 12 + 5.013

Step 1: Write the numbers so that the decimal points are lined up. Write the decimal into the whole number. It will always come to the right of the whole number.

$$\begin{array}{l} \mathbf{12.} \\ \underline{\mathbf{+5.013}} \end{array}$$

continued from previous page

Step 2: Make sure that all the decimals have the same number of digits to the right of the decimal point by adding zeros to the end of shorter decimals.

$$\begin{array}{r} 12.000 \\ \underline{+5.013} \end{array}$$

Step 3: Write the decimal point in the answer so that it lines up with the decimal points in the problem.

$$\begin{array}{r} 12.000 \\ \underline{+5.013} \\ . \quad \end{array}$$

Step 4: Add the decimals just as you would if you were adding whole numbers.

$$\begin{array}{r} 12.000 \\ \underline{+5.013} \\ 17.013 \end{array}$$

The correct answer is 17.013.

PRACTICE

Add the following decimals. You can check your answers at the end of the section.

21. 22.01 + 2.1

22. 3.5 + 6

23. 12.03 + 4.90

24. 1.7 + 4.89

25. 13.4 + 5.67

26. 4.8 + 3.45

27. 3.05 + 0.005

28. 4.601 + 3.01 + 5

29. 4.5 + 5 + 2.09

30. 3.14 + 8 + 2.3

▶ SUBTRACTING DECIMALS

Subtracting decimals is very similar to adding them. Follow these steps to subtract decimals.

Step 1: Write the numbers so that the decimal points are lined up.
Step 2: Make sure that all the decimals have the same number of digits to the right of the decimal point by adding zeros to the end of shorter decimals.
Step 3: Write the decimal point in the answer so that it lines up with the decimal points in the problem.
Step 4: Subtract the decimals just as you would if you were subtracting whole numbers.

Example: 11 – 5.2

Step 1: Write the numbers so that the decimal points are lined up.

```
  11.
– 5.2
```

Step 2: Make sure that all the decimals have the same number of digits to the right of the decimal point by adding zeros to the end of shorter decimals.

```
  11.0
– 5.2
```

Step 3: Write the decimal point in the answer so that it lines up with the decimal points in the problem.

```
  11.0
– 5.2
    .
```

Step 4: Subtract the decimals—don't forget to borrow!

```
 10 10
  11.0
– 5.2
   5.8
```

The correct answer is 5.8.

PRACTICE

Subtract the following decimals. You can check your answers at the end of the section.

31. 5.5 – 2.2

32. 8.3 – 6.3

33. 9.1 – 7.65

34. 3 – 2.8

35. 7.79 – 6.9

36. 4.3 – 2.83

37. 9.003 – 1.2

38. 5.41 – 2.99

39. 10 – 1.999

40. 8 – 4.105

Multiplying and Dividing Decimals

LESSON SUMMARY

In this lesson, you will learn how to multiply and divide decimals.

If you can multiply and divide whole numbers, then you can multiply and divide decimals. The main thing to watch out for is the placement of the decimal point. Placing the decimal point in your answer is just a matter of counting place values.

MULTIPLYING DECIMALS

When you multiply decimals, you begin as if you were working with whole numbers. Then, you come back to the decimal point. When multiplying decimals, follow these steps.

Step 1: Multiply the numbers as if you were working with whole numbers. You can ignore the decimal points for now.
Step 2: Now count the number of decimal places to the right of the decimal point—in each number you are multiplying.

Step 3: Beginning with the digit farthest to the right in your answer, count out the number of decimal places you found in Step 2. Then, write the decimal point to the left of the last digit counted. You should end up with the same number of decimal places in your answer as the sum of decimal places in the numbers you multiplied together.

You don't need to line up the decimal points when multiplying decimals. In fact, you can pretend as if the decimals aren't even there until after you have completed the multiplication. After multiplying the numbers together, then worry about the decimal point. Remember, the number of decimal places in your answer should equal the sum of the decimal places in the numbers with which you began multiplying.

EXAMPLE: How many decimal places will the answer to this problem have?

$$2.8975 \times 1.0104$$

First, count the number of decimal places in each number given. The first number 2.8975 has four decimal places. The second number 1.0104 also has four decimal places. Then, add these two numbers together: 4 + 4 = 8. So, your answer should have eight decimal places.

Let's say you punch 2.8975 × 1.0104 into your calculator. Your calculator might give you the answer: 2.927634. Wait a minute! There aren't eight decimal places to the right of the decimal. There are only six. So, is your calculator wrong? No, it's not. If you multiply out the answer by hand, you'll see that the last two digits are zeros.

```
    2.8975
  x 1.0104
    115900
   289750
+ 289750
2.91763400
```

continued from previous page

As you know, zeros at the end of the decimal do not change its value. Your calculator omitted these zeros in its answer. And you can, too, when you write your final answer.

Example: 0.5 × 0.5

Step 1: Multiply the numbers.

5 × 5 = 25

Step 2: Count the number of decimal places to the right of the decimal point—in each number you are multiplying. Both numbers have one decimal place to the right of decimal point: 1 + 1 = 2.

Step 3: Beginning with the digit farthest to the right in your answer, count out the number of decimal places you found in step 2. That's 2 for this problem. Then, write the decimal point to the left of the last digit counted: 0.25

The answer is 0.25.

When multiplying by a multiple of 10, such as 10, 100, 1,000, and so on, you just have to move the decimal point to the right.

EXAMPLE: 10 × 0.5

Step 1: Count the number of zeros in the multiple of 10. The number 10 has only one zero.

Step 2: Move the decimal point in the decimal you are multiplying to the right the number of zeros you counted in Step 1. You counted one zero in step one, so you move the decimal point over one place in 0.5.

The answer is 5.

EXAMPLE: 100 × 0.5

Step 1: Count the number of zeros in the multiple of 10. The number 100 has two zeros.

continued from previous page

Step 2: Move the decimal point in the decimal you are multiplying to the right the number of zeros you counted in Step 1. You counted two zeros in step one, so you move the decimal point over two places in 0.5.

But there aren't two places to move the decimal point in 0.5, so you add zeros to the right of the 5.

The answer is 50.

PRACTICE

Multiply the following decimals. You can check your answers at the end of the section.

1. 1.5 × 10

2. 0.2 × 0.3

3. 0.92 × 1000

4. 3.4 × 0.1

5. 0.6 × 0.007

6. 4 × 3.12

7. 2.5 × 2.5

8. 0.005 × 0.004

9. 9.01 × 2.01

10. 0.00004 × 4.001

DIVIDING DECIMALS BY WHOLE NUMBERS

Dividing decimals by whole numbers is very similar to dividing whole numbers by whole numbers. The main trick is to set the problem up correctly, so you put the decimal in the correct position. Follow these steps when dividing a decimal by a whole number.

Step 1: Write the decimal point in the answer portion of the division problem. It should go just above the decimal point in the problem.
Step 2: Divide as if you were dividing whole numbers. Ignore the decimal point for the moment.

Example: 4.9 ÷ 7

Step 1: Write the decimal point in the answer portion of the division problem. It should go just above the decimal point in the problem.

$$7\overline{)4.9}$$

Step 2: Divide as if you were dividing whole numbers.

$$\begin{array}{r} 0.7 \\ 7\overline{)4.9} \\ \underline{4\,9} \end{array}$$

The answer is 0.7.

When dividing by a multiple of 10, such as 10, 100, 1,000, and so on, you just have to move the decimal point to the left.

EXAMPLE: 0.5 ÷ 10

Step 1: Count the number of zeros in the multiple of 10: 10 has only one zero.

Step 2: You counted one zero in step one, so you move the decimal point one place to the left: 0.5

0.0.5

Therefore, the answer is 0.05.

continued from previous page

EXAMPLE: 0.5 ÷ 100

Step 1: Count the number of zeros in the multiple of 10. The number 100 has two zeros.

Step 2: You counted two zeros in step one, so you move the decimal point two places to the left. You can add zeros to the left of the number if you run out of places: 00.5

.0 0.5

Therefore, the answer is 0.005.

PRACTICE

Solve the following problems. You can check your answers at the end of the section.

11. 5)14.5

12. 10)59.6

13. 3)51.3

14. 4)600.4

15. 100)56.981

▶ DIVIDING BY DECIMALS

Some problems will require you to divide by a decimal. When the number you are dividing by is a decimal, you should first change it to a whole number. Here's how it works.

Step 1: Make the divisor—the number you are dividing by—a whole number. You do this by moving the decimal point to the right of the number in the ones place.
Step 2: Move the decimal point in the dividend—the number you are dividing into—the same number of places to the right as you did for the divisor.
Step 3: Divide as usual.

Example: $0.03\overline{)1.515}$

Step 1: Make the divisor a whole number.

$0.03\overline{)1.515}$

Step 2: Move the decimal point in the dividend the same number of places to the right.

$3\overline{)1.515}$

Step 3: Divide as usual.

```
    50.5
3 )151.5
 - 15
    01
   - 0
    15
  - 15
     0
```

The answer is 50.5.

What if you are dividing a decimal into a whole number and there aren't enough decimal places to move the decimal to the right? You would follow the same steps, but you would add zeros on to the end of the whole number, so you could move the decimal point the correct number of places to the right.

Example: $0.03\overline{)15}$

Step 1: Make the divisor a whole number.

$0.03\overline{)15}$

Step 2: Move the decimal point in the dividend the same number of places to the right. To do this, you will first have to add the decimal point to the whole number. Remember the decimal point always comes to the right

continued from previous page

of a whole number. Then, add zeros after the decimal point. Now you can move the decimal point over.

$$3\overline{)15.00}$$

Add the decimal point here.

Step 3: Divide as usual.

$$\begin{array}{r} 500. \\ 3\overline{)1500.} \\ \underline{15} \\ 000 \end{array}$$

The answer is 500.

Remember that adding zeros to the right of a decimal does not change its value. So anytime you need to move a decimal point more places to the right, just add zeros.

PRACTICE

Solve the following problems. You can check your answers at the end of the section.

16. 43.5 ÷ 5

17. 57.6 ÷ 3

18. 89.2 ÷ 10

19. 192.6 ÷ 6

20. 2.59104 ÷ 1000

21. 55 ÷ 0.2

22. 3.36 ÷ 1.6

23. 32.24 ÷ 5.2

24. 16.16 ÷ 0.04

25. 32 ÷ 0.004

There's an easy way to check your division: just multiply your answer by the divisor (the number you are dividing by).

Example: $4\overline{)35.6}$

First, solve the problem.

$$
\begin{array}{r}
8.9 \\
4\overline{)35.6} \\
\underline{32} \\
36 \\
\underline{36} \\
0
\end{array}
$$

Then, check your answer with multiplication.

$4 \times 8.9 = 35.6$

Your answer is correct.

Real World Problems

These problems apply the skills you've learned in Section 3 to everyday situations. As you work through these problems, you'll see that the skills you've learned in this section aren't only important for math tests. They are important skills for ordinary questions that come up every day. You can check your answers at the end of the section.

Amounts of money are represented to the hundredths place. For example, we write one and one-half dollars as $1.50, not as 1.5. When you use a calculator to work with amounts of money, remember to write your final answer to the hundredths place. This means two things:

- **You should add a zero to a decimal that goes only to the tenths place. So you would write $1.2 as $1.20.**
- **You should round decimals that go beyond the hundredths place to the nearest hundredths place. So you would write $1.278 as $1.28.**

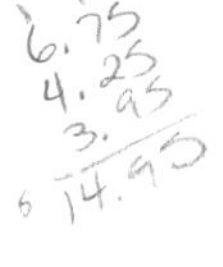

1. Kahn bought a book online and had it gift wrapped for her nephew's birthday. If the book cost $6.75, the gift wrapping cost $4.25, and the shipping cost $3.95, how much did the gift cost?

2. Sameer jogged 1.2 miles on Saturday, 2.5 miles on Monday, and 1.75 miles on Thursday. How many miles did he jog in all?

3. Doak has $45. If he buys a belt for $12.99 and a CD for $15.50, how much money will he have left over to pay for the sales tax on these items?

4. Look at the following price list for computer games at a local computer software store. Then, answer the questions.

COMPUTER GAME	RETAIL PRICE	SALE PRICE
Golf World	$22.99	$19.99
Mega Hits	$23.25	$20.25
Designer Fun	$29.67	$25.85

a. Valerie wants to buy Mega Hits and Designer Fun. What will the two games cost at their full retail prices?

b. How much money can Valerie save by waiting until Mega Hits and Designer Fun go on sale?

c. If Valerie has $47.50, does she have enough money to buy both games? Explain your answer.

d. Valerie's grandparents are sending her $50 for her birthday. If she uses the money to buy the two computer games on sale, how much of the $50 will she have left over?

5. Yuri works in the school library. He needs to put away books with these call numbers in sequential order: 513.26, 513.59, 513.7, 513.514. In which order should Yuri place these books on the shelf?

6. Roberta bought four shirts at $12.95 each and a hat for $7.85. What was the total cost of the shirts and the hat before sales tax?

7. The chart below shows the times that four swimmers had in a race. By how much did the fastest swimmer beat the slowest swimmer?

SWIMMER	TIME (IN SECONDS)
Molly	38.51
Jeff	39.23
Asta	37.95
Risa	37.89

8. Rachel has a vegetable garden in her backyard. The area of the garden is 423.7 square feet. If she divides the garden into 5 equal plots, how large will each plot be?

9. Estrella wants to buy a rug for her bedroom. The dimensions of the room are 15.6 feet by 27.75 feet. What is the largest rug size she can fit on the floor of her room?

10. Malita had $5.48 of her own money and borrowed some more money from her mother. Then she went shopping. She bought a CD for $15.71 as a birthday gift for her brother. After that Malita spent $4.67 for lunch and had $3.60 left. How much money did she borrow from her mother?

SECTION

3

Answers & Explanations

LESSON 7

1. five hundredths
2. two tenths
3. twenty-two hundredths
4. one and two hundred thirty-four thousandths
5. twenty-five and eighty-nine hundredths
6. thirteen and nine tenths or thirteen and ninety hundredths
7. one hundred and one thousandth
8. forty-five and one hundred two thousandths
9. seven and eight thousand thirty-five ten thousandths
10. two and three hundredths or two and thirty thousandths
11. 0.13
12. 0.4
13. 0.025
14. 0.06
15. 0.006
16. 3.028
17. 18.12
18. 0.0012
19. 25.605
20. 65.0105

21. To solve this problem correctly, you must be sure that the decimal points are lined up correctly. Then, add a zero to the end of 2.1 so that the numbers have the same number of digits. Now you are ready to add:

$$\begin{array}{r} 22.01 \\ \underline{+2.10} \\ 24.11 \end{array}$$

22. To solve this problem correctly, you must be sure that the decimal points are lined up correctly. Remember that the decimal point for the whole number 6 comes just to the right of the number. Then, add a zero to the end of 6.0 so that the numbers have the same number of digits. Finally, add:

$$\begin{array}{r} 3.5 \\ \underline{+6.0} \\ 9.5 \end{array}$$

23. To solve this problem correctly, you must be sure that the decimal points are lined up correctly. Then, add:

$$\begin{array}{r} 12.03 \\ \underline{+4.90} \\ 16.93 \end{array}$$

24. To solve this problem correctly, you must be sure that the decimal points are lined up correctly. Then, add:

$$\begin{array}{r} 1.70 \\ \underline{+4.89} \\ 6.59 \end{array}$$

25. To solve this problem correctly, you must be sure that the decimal points are lined up correctly. Then, add:

$$\begin{array}{r} 13.40 \\ \underline{+\ 5.67} \\ 19.07 \end{array}$$

26. To solve this problem correctly, you must be sure that the decimal points are lined up correctly. Then, add:

$$\begin{array}{r} 4.80 \\ \underline{+\ 3.45} \\ 8.25 \end{array}$$

27. To solve this problem correctly, you must be sure that the decimal points are lined up correctly. Then, add:

$$\begin{array}{r} 3.050 \\ \underline{+0.005} \\ 3.055 \end{array}$$

28. To solve this problem correctly, you must be sure that the decimal points are lined up correctly. Then, add:

$$\begin{array}{r} 4.601 \\ 3.010 \\ \underline{+5.000} \\ 12.611 \end{array}$$

29. To solve this problem correctly, you must be sure that the decimal points are lined up correctly. Then, add:

$$\begin{array}{r} 4.50 \\ 5.00 \\ \underline{+2.09} \\ 11.59 \end{array}$$

30. To solve this problem correctly, you must be sure that the decimal points are lined up correctly. Then, add:

$$\begin{array}{r} 3.14 \\ 8.00 \\ \underline{+2.30} \\ 13.44 \end{array}$$

31. To solve this problem correctly, you must be sure that the decimal points are lined up correctly. Then, subtract:

$$\begin{array}{r} 5.5 \\ -\underline{2.2} \\ 3.3 \end{array}$$

32. To solve this problem correctly, you must be sure that the decimal points are lined up correctly. Then, subtract:

$$\begin{array}{r} 8.3 \\ -\underline{6.3} \\ 2.0 \end{array}$$

33. To solve this problem correctly, you must be sure that the decimal points are lined up correctly. Then, subtract:

$$\begin{array}{r} 9.10 \\ -\underline{7.65} \\ 1.45 \end{array}$$

34. To solve this problem correctly, you must be sure that the decimal points are lined up correctly. Then, subtract:

$$\begin{array}{r} 3.0 \\ -\underline{2.8} \\ 0.2 \end{array}$$

35. To solve this problem correctly, you must be sure that the decimal points are lined up correctly. Then, subtract:

$$\begin{array}{r} 7.79 \\ -\underline{6.90} \\ 0.89 \end{array}$$

36. To solve this problem correctly, you must be sure that the decimal points are lined up correctly. Then, subtract:

$$\begin{array}{r} 4.30 \\ -\underline{2.83} \\ 1.47 \end{array}$$

37. To solve this problem correctly, you must be sure that the decimal points are lined up correctly. Then, subtract:

$$\begin{array}{r} 9.003 \\ -\underline{1.200} \\ 7.803 \end{array}$$

38. To solve this problem correctly, you must be sure that the decimal points are lined up correctly. Then, subtract:

$$\begin{array}{r} 5.41 \\ -\underline{2.99} \\ 2.42 \end{array}$$

39. To solve this problem correctly, you must be sure that the decimal points are lined up correctly. Then, subtract:

$$\begin{array}{r} 10.000 \\ -\underline{1.999} \\ 8.001 \end{array}$$

40. To solve this problem correctly, you must be sure that the decimal points are lined up correctly. Then, subtract:

$$\begin{array}{r} 8.000 \\ -\underline{4.105} \\ 3.895 \end{array}$$

LESSON 8

1. Begin by multiplying the numbers together: 15 × 10 = 150. Now you have to insert the decimal point in the correct place. Count the number of places to the right of the decimal point in each number you are multiplying. You have a total of one decimal place (1 place in 1.5 and none in 10), so you move the decimal place one place to the left, and your answer is 15.

Use a shortcut to solve this problem. First, count the number of zeros in the multiple of 10. The number 10 has only one zero. Then, move the decimal point in 1.5 to the right one digit. The answer is 15.

2. Begin by multiplying the numbers: 2 × 3 = 6. Now you have to insert the decimal point in the correct place. Count the number of places to the right of the decimal point in each number you are multiplying. You have a total of two decimal places (one place in .2 and one place in .3), so you move the decimal two places to the left, and your answer is 0.06.

.06

3.

Use the shortcut to solve this problem. First, count the number of zeros in the multiple of 10. The number 1,000 has three zeros. Then, move the decimal point in 0.92 to the right three digits. The answer is 920.

4. Begin by multiplying the numbers: 3.4 × .1 = 34. Now you have to insert the decimal point in the correct place. Count the number of places to the right of the decimal point in each number you are multiplying. You have a total of two decimal places (one place in 3.4 and one place in .1), so you move the decimal point over two places to the left in your answer to get 0.34.

.34

5. Begin by multiplying the numbers: 6 × 7 = 42. Now you have to insert the decimal point in the correct place. Count the number of places to the right of the decimal point in each number you are multiplying. You have a total of four decimal places (one place in 0.6 and three places in 0.007), so you move the decimal point four places to the left to get your answer: 0.0042.

.0042

6. Begin by multiplying the numbers: 4 × 312 = 1,248. Now you have to insert the decimal point in the correct place. Count the number of places to the right of the decimal point in each number you are multiplying. You have a total of two decimal places (two places 3.12 and none in 4), so you move the decimal point two places to the left to get your answer: 12.48.

12.48

7. Begin by multiplying the numbers: 25 × 25 = 625. Now you have to insert the decimal point in the correct place. Count the number of places to the right of the decimal point in each number you are multiplying. You have a total of two decimal places (one place in each 2.5), so you move the decimal point two places to the left to get your answer: 6.25.

6.25

8. Begin by multiplying the numbers: 5 × 4 = 20. Now you have to insert the decimal point in the correct place. Count the number of places to the right of the decimal point in each number you are multiplying. You have a total of six decimal places (three places .005 and three in .004), so you move the decimal point six places to the left to get your answer: 0.00002.

.000020

9. Begin by multiplying the numbers: 901 × 201 = 181,101. Now you have to insert the decimal point in the correct place. Count the number of places to the right of the decimal point in each number you are multiplying. You have a total of four decimal places (two places 9.01 and two places in 2.01), so you move the decimal point four places to the left to get your answer: 18.1101.

18.1101

10. Begin by multiplying the numbers: 4 × 4,001 = 16,004. Then, count the number of decimal places to the right of the decimal point—in each number you are multiplying. The decimal point is five places to the left in 0.00004 and three places to the left in 4.001, so you move the decimal point over eight places in your answer to get 0.00016004.

.00016004

11. First, you need to set up the division problem. Begin by writing the decimal point in the answer portion of the division problem. It should go just above the decimal point in the problem. Then, divide as if you were dividing whole numbers.

$$\begin{array}{r} 2.9 \\ 5\overline{)14.5} \\ -\underline{10} \\ 45 \\ -\underline{45} \\ 0 \end{array}$$

Your final answer is 2.9.

12.

Use the shortcut to solve this problem. First, count the number of zeros in the multiple of 10. The number 10 has only one zero. Then, move the decimal point in 59.6 to the left one digit. Your final answer is 5.96.

13. First, you need to set up the division problem. Begin by writing the decimal point in the answer portion of the division problem. It should go just above the decimal point in the problem. Then, divide as if you were dividing whole numbers.

$$\begin{array}{r} 17.1 \\ 3\overline{)51.3} \\ -\underline{513} \\ 0 \end{array}$$

Your final answer is 17.1

14. First, you need to set up the division problem. Begin by writing the decimal point in the answer portion of the division problem. It should go just above the decimal point in the problem. Then, divide as if you were dividing whole numbers.

$$\begin{array}{r} 150.1 \\ 4\overline{)600.4} \\ -\underline{6004} \\ 0 \end{array}$$

Your final answer is 150.1.

15.

Use the shortcut to solve this problem. First, count the number of zeros in the multiple of 10. The number 100 has two zeros. Then, move the decimal point in 56.981 to the left two digits. Your final answer is 0.56981.

16. First, you need to set up the division problem. Begin by writing the decimal point in the answer portion of the division problem. It should go just above the decimal point in the problem. Now you are ready to divide:

$$\begin{array}{r} 8.7 \\ 5\overline{)43.5} \\ -\underline{40} \\ 35 \\ -\underline{35} \\ 0 \end{array}$$

Your final answer is 8.7.

17. First, you need to set up the division problem. Begin by writing the decimal point in the answer portion of the division problem. It should go just above the decimal point in the problem. Now you are ready to divide:

$$\begin{array}{r} 19.2 \\ 3\overline{)57.6} \\ -3 \\ \hline 27 \\ -27 \\ \hline 06 \end{array}$$

Your final answer is 19.2

18.

Use the shortcut to solve this problem. First, count the number of zeros in the multiple of 10. The number 10 has one zero. Then, move the decimal point in 89.2 to the left one digit. Your final answer is 8.92.

19. First, you need to set up the division problem. Begin by writing the decimal point in the answer portion of the division problem. It should go just above the decimal point in the problem. Now you are ready to divide:

$$\begin{array}{r} 32.1 \\ 6\overline{)192.6} \\ -18 \\ \hline 12 \\ -12 \\ \hline 06 \end{array}$$

Your final answer is 32.1.

20.

Use the shortcut to solve this problem. First, count the number of zeros in the multiple of 10. The number 1,000 has three zeros. Then, move the decimal point in 2.59104 to the left three digits. Your final answer is 0.002591.

21. First, set up the division problem. Begin by making the divisor a whole number, so 0.2 becomes 2. Then, move the decimal point in the dividend the same number of places to the right. To do this, you will first have to add the decimal point to the whole number. Then, you can add zeros after the decimal point, so that your dividend is now 55.0. Now you are ready to divide:

```
  27.5
2)55.0
 – 4
  15
 – 14
   10
  – 10
    0
```

Your final answer is 27.5

22. First, set up the division problem. Begin by making the divisor a whole number, so 1.6 becomes 16. Then, move the decimal point in the dividend the same number of places to the right, so that your dividend becomes 33.6. Now you are ready to divide:

```
     2.1
16)33.6
  – 32
     16
   – 16
      0
```

Your final answer is 2.1

23. First, set up the division problem. Begin by making the divisor a whole number, so 5.2 becomes 52. Then, move the decimal point in the dividend the same number of places to the right, so that your dividend becomes 322.4. Now you are ready to divide:

```
      6.2
52)322.4
  – 312
     104
   – 104
       0
```

Your final answer is 6.2.

24. First, set up the division problem. Begin by making the divisor a whole number, so 0.04 becomes 4. Then, move the decimal point in the dividend the same number of places to the right, so that your dividend becomes 1,616. Now you are ready to divide:

```
   404.
4)1,616.
 - 16
    01
   - 0
    16
  - 16
     0
```

Your final answer is 404.

25. First, set up the division problem. Begin by making the divisor a whole number, so 0.004 becomes 4. Then, move the decimal point in the dividend the same number of places to the right. To do this, you will first have to add the decimal point to the whole number. Then, you can add zeros after the decimal point, so that your dividend is now 32,000. Now you are ready to divide:

```
   8,000.
4)32,000.
 - 32
   0000
```

Your final answer is 8,000.

REAL WORLD PROBLEMS

1. Set the problem up as follows and add: $6.75 + $4.25 + $3.95 = $14.95. Thus, the total price of the gift was $14.95.

2. Set the problem up as follows and add: 1.2 miles + 2.5 miles + 1.75 miles = 5.45 miles. So, Sameer jogged 5.45 miles in all.

3. Set the problem up as follows: $45 – ($12.99 + $15.50) = $16.51. Doak will have $16.51 left over to pay for sales tax.

4. a. Set the problem up as follows and add: $23.25 + $29.67 = $52.92. At retail cost, the two games will cost $52.92.

b. Set the problem up as follows: $52.92 – ($20.25 + $25.85) = $6.82. By waiting for a sale, she can save $6.82.

c. Based on the calculations in **a** and **b** above, she can only buy the two games if she buys them at the sale price and the tax is included.

d. Set the problem up as follows: $50 – ($20.25 + $25.85) = $3.90. She will have $3.90 left over.

5. Yuri should order the books in sequential order: 513.26, 513.514, 513.59, 513.7.

6. Set the problem up as follows: 4 × \$12.95 + \$7.85 = \$59.65. The total cost of the purchase was \$59.65.

7. Set the problem up as follows and subtract: 39.23 seconds – 37.89 seconds = 1.34 seconds. The fastest swimmer beat the slowest swimmer by 1.34 seconds.

8. Set the problem up as follows and divide: 423.7 square feet ÷ 5 = 84.74 square feet. Each plot will be 84.74 square feet.

9. Set the problem up as follows and multiply: 15.6 feet × 27.75 feet = 432.9 square feet. The rug needs to be 432.9 square feet or smaller.

10. Set the problem up as follows: \$15.71 + \$4.67 + \$3.60 – \$5.48 = \$18.50. Malita borrowed \$18.50 from her mother.

SECTION

4

Percentages

NOW THAT YOU know all about decimals, it's time to learn about percentages. Percents are just *hundredths*. We've already seen *hundredths* when we addressed fractions and decimals. In this section, we will review how to express given percentages as both fractions and decimals. You'll find that you come into contact with percentages every day with grades, sales tax, tips, and discounts.

Converting Percents, Decimals, and Fractions

LESSON SUMMARY

So far in this book, you've learned how to work with parts of a whole that are represented by fractions and decimals. In this lesson, you will begin working with another way of writing parts of a whole—percents. You will learn what percents are and how to convert them to decimals and fractions. You will also learn how to convert decimals and fractions to percents.

Percents are everywhere you look. Go to the mall, and you'll see plenty of signs announcing "20% off" or "Take an additional 30% off." Packages at the supermarket regularly claim to include "30% more free." Even your grades at school are probably percents!

WHAT ARE PERCENTS?

Like fractions and decimals, *percent* is another way to represent the parts of a whole. Notice that percents are written with the *percent sign* after a number: 10%, 25%, 30%, 50%, 99%, and so on. The percent sign represents the words "out of 100 parts" or "per 100 parts."

Recall that fractions represent the parts of a whole that is divided into any number of equal parts. So, you can find fractions with any whole number in the denominator: $\frac{1}{2}$, $\frac{2}{3}$, $\frac{4}{5}$, $\frac{6}{10}$, $\frac{15}{212}$, and so on. Decimals represent the parts of a whole that is divided into either 10, 100, 1,000, or another multiple of 10 equal parts. Percents, on the other hand, always represent a whole that is divided into 100 equal parts. That means that percents can be written as fractions with 100 in the denominator and decimals written to the hundredths place.

CONVERTING PERCENTS TO DECIMALS

Changing percents to decimals is as simple as moving the decimal point two digits to the left. Here are the basic steps.

Step 1: Drop the percent sign.

Step 2: Add a decimal point if there isn't already one. Remember that even when it's not written in, whole numbers are followed by a decimal point.

Step 3: Move the decimal point two places to the left.

Example: Convert 25% to a decimal.

Step 1: Drop the percent sign.

So 25% becomes 25.

Step 2: Add a decimal point.

25.

Step 3: Move the decimal point two places to the left.

0.25

So 25% = 0.25

Example: Convert 22.5% to a decimal.

Step 1: Drop the percent sign.

22.5

Step 2: There is already a decimal point in place, so you can skip this step.

Step 3: Move the decimal point two places to the left.

0.225

So 22.5% = 0.225

PRACTICE

Convert these percents to decimals. You can check your answers at the end of the section.

1. 25%

2. 12%

3. 50%

4. 62.5%

5. 4.2%

6. 0.2%

7. 125%

8. 100%

9. 128.9%

10. 2,000%

▶ CONVERTING DECIMALS TO PERCENTS

Changing decimals to percents is the opposite of what you've just done. When you change a decimal to a percent, you move the decimal point two digits to the right. Here are the basic steps.

Step 1: Move the decimal point two places to the right. If there aren't enough digits to move the decimal point over two places, add zeros.

Step 2: Add a percent sign after the number.

Example: Convert 0.15 to a percent.

Step 1: Move the decimal point two places to the right.

So 0.15 becomes 15.

Step 2: Add a percent sign after the number.

15%

Thus, 0.15 = 15%.

Example: Convert 7.9 to a percent.

Step 1: Move the decimal point two places to the right. Add zeros as needed.
So 7.9 becomes 790.

Step 2: Add a percent sign after the number.
790%

Thus, 7.9 = 790%.

PRACTICE

Convert these decimals to percents. You can check your answers at the end of the section.

11. 0.40

12. 0.75

13. 0.625

14. 0.29

15. 0.33

16. 1.56

17. 2.0

18. 6.5

19. 3.56

20. 8

▶ CONVERTING PERCENTS TO FRACTIONS

To change a percent to a fraction, you write the percent over 100. Don't forget to reduce the fraction to lowest terms as you would any other fraction. Here are the steps to follow.

Step 1: Drop the percent sign.

Step 2: Write the number as a fraction over 100.

Step 3: Write improper fractions as mixed numbers. Reduce the fraction to lowest terms.

Example: Convert 15% to a fraction.

Step 1: Drop the percent sign.
15% becomes 15.

Step 2: Write the number as a fraction over 100.

$\frac{15}{100}$

Step 3: Reduce the fraction to lowest terms. Both 15 and 100 can be divided by 5.

$\frac{15 \div 5}{100 \div 5} = \frac{3}{20}$

So, 15% = $\frac{3}{20}$.

Example: Convert 150% to a fraction.

Step 1: Drop the percent sign.

150% becomes 150.

Step 2: Write the number as a fraction over 100.

$\frac{150}{100}$

Step 3: Write the improper fraction as a mixed number, and reduce the fraction to lowest terms.

$\frac{150 \div 50}{100 \div 50} = \frac{3}{2} = 1\frac{1}{2}$

So, 150% = $1\frac{1}{2}$.

What if the percent already has a fraction in it?

EXAMPLE: Convert $15\frac{1}{2}$% to a fraction.

Step 1: Drop the percent sign.

$15\frac{1}{2}$

Step 2: Write the number as a fraction over 100.

$\frac{15\frac{1}{2}}{100}$

Step 3: Remember that the bar in a fraction means to divide.

So you can rewrite this problem as a division problem. $15\frac{1}{2} \div 100$

Change the mixed number $15\frac{1}{2}$ to an improper fraction. $\frac{31}{2} \div 100$

Invert the second fraction and multiply. $\frac{31}{2} \times \frac{1}{100} = \frac{31}{200}$

Step 4: Since the fraction is already in lowest terms, you're done.

So, $15\frac{1}{2}$% = $\frac{31}{200}$.

PRACTICE

Convert these percents to fractions. You can check your answers at the end of the section.

21. 16%

22. 5%

23. 25%

24. 80%

25. 34%

26. 10%

27. 89%

28. 3%

29. $87\frac{1}{2}$%

30. $16\frac{2}{3}$%

▶ CONVERTING FRACTIONS TO PERCENTS

There are two basic ways to convert fractions to percents. You should try both ways, and see which one works best for you.

Method 1

Step 1: Divide the numerator by the denominator.
Step 2: Multiply by 100. (This is the same as moving the decimal point two digits to the right.)
Step 3: Add a percent sign.

Method 2

Step 1: Multiply the fraction by $\frac{100}{1}$.
Step 2: Write the product as either a whole or a mixed number.
Step 3: Add a percent sign.

Example: Change $\frac{2}{5}$ to a percent.

Method 1

Step 1: Divide the numerator by the denominator.

$$5\overline{)2.0}\quad = 0.4,\quad -20$$

Step 2: Multiply by 100.

$0.4 \times 100 = 40$

Step 3: Add a percent sign.

40%

Method 2

Step 1: Multiply the fraction by $\frac{100}{1}$.

$\frac{2}{5} \times \frac{100}{1} = \frac{200}{5} = \frac{40}{1}$

Step 2: Write the product as either a whole or a mixed number.

40

Step 3: Add a percent sign.

40%

To convert a mixed number to a percent, first change it to an improper fraction. Then, follow either Method 1 or Method 2 above.

EXAMPLE: Convert $2\frac{1}{2}$ to a percent.

Step 1: Change the mixed number to an improper fraction.

$2\frac{1}{2} = \frac{5}{2}$

Step 2: Multiply the fraction by $\frac{100}{1}$.

$\frac{5}{2} \times \frac{100}{1} = \frac{500}{2} = \frac{250}{1}$

Step 3: Write the product as either a whole or a mixed number.

250

Step 4: Add a percent sign.

250%

So, $2\frac{1}{2}$ = 250%.

PRACTICE

Convert these fractions to percents. You can check your answers at the end of the section.

31. $\frac{3}{4}$

32. $\frac{1}{2}$

33. $\frac{3}{5}$

34. $\frac{1}{4}$

35. $\frac{3}{50}$

36. $\frac{1}{8}$

37. $\frac{7}{10}$

38. $\frac{17}{20}$

39. $\frac{19}{25}$

40. $\frac{18}{5}$

41. $3\frac{1}{4}$

42. $9\frac{4}{5}$

TEST TAKING TIP

Although you can always convert between percents, decimals, and fractions using the described methods, it's a good idea to know common percent, decimal, and fraction equivalents for standardized tests. Knowing them in advance can save you valuable time on a timed test. Besides, working with a value in one form is often easier than working with it in another form. Knowing the equivalents can help you see the easier route faster. Here are some common equivalents you might want to learn.

continued from previous page

EQUIVALENTS TO KNOW

Percent	Decimal	Fraction
1%	0.01	$\frac{1}{100}$
5%	0.05	$\frac{5}{100}$
10%	0.1	$\frac{1}{10}$
12.5%	0.125	$\frac{1}{8}$
20%	0.2	$\frac{1}{5}$
25%	0.25	$\frac{1}{4}$
$33\frac{1}{3}$%	0.33	$\frac{1}{3}$
40%	0.40	$\frac{2}{5}$
50%	0.5	$\frac{1}{2}$
$66\frac{2}{3}$%	0.67	$\frac{2}{3}$
75%	0.75	$\frac{3}{4}$
80%	0.80	$\frac{4}{5}$
90%	0.90	$\frac{9}{10}$
100%	1.00	$1 = \frac{1}{1}$

Solving Percent Problems

LESSON SUMMARY

In this lesson, you will use percents to solve math problems. You will learn about three basic kinds of percent problems and how to solve them.

Percent problems ask you to find one of three things: the part, the whole, or the percent. Here's how these three elements are related to one another.

Whole × *Percent* = *Part*

This is called an *equation*, or a kind of math sentence. It tells how different elements are related to one another. You can use this equation to find any one of the elements that might be missing.

It's often easier to multiply a percent if you first convert it to a decimal.

EXAMPLE: 10% × 12
Convert 10% to a decimal: .10, or 0.1
Then, multiply: 0.1 × 12 = 1.2

▶ FINDING A PART OF A WHOLE

Often you will be asked to find a part of a whole. In these problems, you are given a whole and a percent, and are asked to find the part represented by the percent of the whole. Let's go through some examples.

Example: What is 30% of 60?

Step 1: Begin by figuring out what you know from the problem and what you're looking for.

You have the percent: 30%.

You have the whole: 60.

You are looking for the part.

Step 2: Then, use the equation to solve the problem: Whole Percent = Part

Plug in the pieces of the equation that you know: 60 × 30% = Part

Step 3: Convert the percent to a decimal to make your multiplication easier: 60 × 0.30 = Part

Step 4: Solve: 60 × 0.30 = 18.

So, 30% of 60 is 18.

Example: Only 10% of Mrs. Cunningham's class got an A on the last test. There are 20 students in the class. How many students got an A on the last test?

Step 1: Begin by figuring out what you know from the problem and what you're looking for.

You have the percent: 10%.

You have the whole: 20 students.

You are looking for the part—the part of the class that got an A on the last test.

Step 2: Then, use the equation to solve the problem: Whole × Percent = Part

Plug in the pieces of the equation that you know: 20 × 10% = Part

Step 3: Convert the percent to a decimal to make your multiplication easier: 20 × 0.10 = Part

Step 4: Solve: 20 × 0.10 = 2.

So, two students received an A on Mrs. Cunningham's last test.

PRACTICE

Solve each problem. You can check your answers at the end of the section.

1. What is 1% of 34?

2. What is 10% of 52?

3. What is 0.5% of 30?

4. What is 100% of 99?

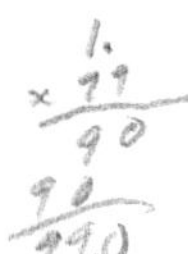

5. What is 25% of 100?

6. What is 20% of 70?

7. What is 90% of 10?

8. What is 80% of 50?

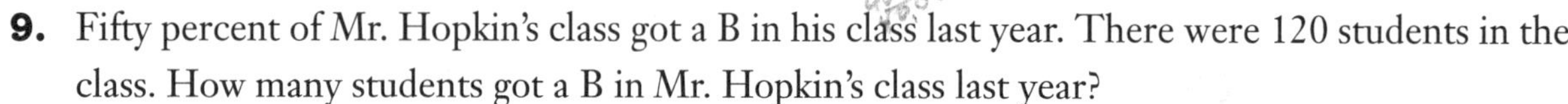

9. Fifty percent of Mr. Hopkin's class got a B in his class last year. There were 120 students in the class. How many students got a B in Mr. Hopkin's class last year?

10. Lan's class sold candy bars to raise money for the school French Club. Lan sold 40% of the 500 candy bars that were sold. How many candy bars did Lan sell?

FINDING A PERCENT

In the following types of problems, you will be given the part and the whole. Your task is to determine what percent the part is of the whole. Remember that a percent is just a fraction written over 100. You can solve these types of problems by writing the part over the whole and converting the fraction to a percent.

Example: 10 is what percent of 200?

Step 1: Begin by figuring out what you know from the problem and what you're looking for.

You have the part: 10.

You have the whole: 200.

You are looking for the percent.

Step 2: Write a fraction of the part over the whole.

$\frac{10}{200}$

Step 3: Convert the fraction to a percent. Remember there are two methods for converting fractions to percents. Use either method. Method 1 is shown below.

$10 \div 200 = 0.05$

$0.05 \times 100 = 5$

5%

So, 10 is 5% of 200.

Let's look at the equation from earlier in the lesson again.

$$Whole \times Percent = Part$$

If you have already learned some algebra, you know that the elements in an equation can be rearranged in order to solve for one element in the equation. Thus, another way to approach this type of problem is to go back to the original equation and rearrange the elements, so that the equation is set to solve for the percent. In this case, the equation would look like this:

$$Percent = \frac{Part}{Whole}$$

Notice that when you set up a fraction with the part over the whole, you are really just solving this equation.

Example: There are 500 people in Sandra's 8th grade class. Fifty people in her class were chosen to go to Washington, DC, for a field trip. What percent of Sandra's class was chosen to go on the field trip?

Step 1: Begin by figuring out what you know from the problem and what you're looking for.

You have the part: 50.

You have the whole: 500.

You are looking for the percent.

Step 2: Write a fraction of the part over the whole.

$\frac{50}{500}$

Step 3: Convert the fraction to a percent. Remember there are two methods for converting fractions to percents. Use either method. Method 1 is shown below.

$50 \div 500 = 0.1$

$0.1 \times 100 = 10$

10%

So, 10% of Sandra's class was chosen to go on the field trip.

Sometimes, this type of problem will ask you to find a percent change. The problem will give you one part of the whole. Your task is to calculate the part of the whole represented by the *difference* between the whole and the part given.

EXAMPLE: Last year, Sasha could run one mile in 10 minutes. This year, he can run a mile in 8 minutes. By what percent did his timing improve?

Step 1: Begin by figuring out what you know from the problem and what you're looking for.

You have the part: 10 minutes – 8 minutes = 2 minutes

You also have the whole: 10 minutes.

You are looking for the percent.

Step 2: Write a fraction of the part over the whole.

$$\frac{2 \text{ minutes}}{10 \text{ minutes}}$$

Step 3: Convert the fraction to a percent.

$2 \div 10 = 0.2$

$0.2 \times 100 = 20$

20%

So, Sasha's running time improved by 20% since last year.

When you see these types of phrases, you are probably being asked to calculate the part of the whole represented by the *difference* between the whole and the part given.

- Find the percent change.
- Find the percent increase.
- Find the percent decrease.
- By what percent did it improve?
- By what percent did it go down?
- By what percent did it go up?

PRACTICE

Solve each problem. You can check your answers at the end of the section.

11. 10 is what percent of 50?

12. 20 is what percent of 80?

13. 5 is what percent of 20?

14. 6 is what percent of 6?

15. 9 is what percent of 18?

16. 1 is what percent of 50?

17. 22 is what percent of 100?

18. There are 30 people in Mac's 7th grade class. Three people in his class were awarded medals for outstanding achievement. What percent of Mac's class received one of these awards?

19. Only 12 students are selected to be cheerleaders each year at Mickey's school. If 60 students try out for cheerleader, what percent of those trying out are selected to be cheerleaders?

20. Kim's average in History class went down from 100 to 95. By what percent did his average go down?

Sometimes the problem will not directly tell you the whole amount. Instead, you will be given enough information to calculate the whole on your own. Here's an example.

EXAMPLE: Priyanka made a beaded bracelet using 10 red beads, 5 turquoise beads, 8 yellow beads, and 2 white beads. What percent of the bracelet do the red beads make up?

Step 1: Begin by figuring out what you know from the problem and what you're looking for.

You have the parts:

10 red beads
5 turquoise beads
8 yellow beads
2 white beads

continued from previous page

You also have the whole: 10 + 5 + 8 + 2 = 25 beads.
You are looking for the percent.

Step 2: Write a fraction of the part over the whole.

$\frac{\text{10 red beads}}{\text{25 total beads}}$

Step 3: Convert the fraction to a percent.

10 ÷ 25 = 0.4
0.4 × 100 = 40
40%
So, 40% of the beads are red.

FINDING THE WHOLE

In the following types of problems, you will be given the part and the percent. Your task is to determine the whole. You can solve these types of problems by writing the part over the percent and dividing.

$$Whole = \frac{Part}{Percent}$$

Example: 45 is 75% percent of what number?

Step 1: Begin by figuring out what you know from the problem and what you're looking for.
You have the part: 45
You also have the percent: 75%
You are looking for the whole.

Step 2: Write the part over the percent.

$\frac{45}{75\%}$

Step 3: Convert the percent to a fraction to make your division easier: $45 \div \frac{75}{100}$ = Whole

Step 4: Solve:

$45 \div \frac{75}{100}$

$\frac{45}{1} \times \frac{100}{75} = 60$

So, 45 is 75% of 60.

Example: Five students on the track team qualified to go to the State competition. This represents 20% of the track team. How many students are on the track team?

Step 1: Begin by figuring out what you know from the problem and what you're looking for.

You have the part: 5 students

You also have the percent: 20%

You are looking for the whole.

Step 2: Write the part over the percent.

$\frac{5}{20\%}$

Step 3: Convert the percent to a fraction to make your division easier: $5 \div \frac{20}{100}$ = Whole

Step 4: Solve:

$5 \div \frac{20}{100}$

$\frac{5}{1} \times \frac{100}{20} = 25$

So, there are 25 students on the track team.

PRACTICE

Solve these problems. You can check your answers at the end of the section.

21. 20 is 20% percent of what number?

22. 30 is 75% percent of what number?

23. 6 is 50% percent of what number?

24. 75 is 75% percent of what number?

25. 20 is 2% percent of what number?

26. 40% percent of what number is 100?

27. 10% percent of what number is 5?

28. 80% of what number is 120?

29. 200% of what number is 50?

30. One student on the track team sprained her ankle last week. This represents 2% of the track team. How many students are on the track team?

Instead of rearranging the percent equation for each type of problem, you can use it to set up a proportion like the one shown below.

$$\frac{Part}{Whole} = \frac{\%}{100}$$

You can use this proportion to solve all three types of percent problems. You cross multiply to solve each time. To *cross multiply,* multiply the numerator (or top number) of the first fraction by the denominator (or bottom number) of the second fraction and the denominator of the first fraction by the numerator of the second fraction. The result will look like this:

$$Part \times 100 = Whole \times \%$$

Real World Problems

These problems apply the skills you've learned in Section 4 to everyday situations. As you work through these problems, you'll see that the skills you've learned in this section aren't only important for math tests. They are important skills for ordinary questions that come up every day. You can check your answers at the end of the section.

Before calculating the answer to a word problem, first determine what pieces of information are known and what pieces of information the question expects you to calculate.

1. Jackie bought a dress on sale. The original price for the dress was $50. She got 25% off the original price. How much money did she save?

2. Estrella's car was damaged in an accident. Her insurance company will pay for 90% of the cost of the repairs. If the repairs cost $800, how much money does Estrella need to pay from her own funds?

3. The label on a shampoo bottle says "33% free." If the bottle contains 20 ounces, how much of the shampoo is supposed to be free?

4. Marguerite got 80% of the items on a test correct. If she got 56 items correct, what was the total number of items on the test?

5. Sheila has a hard time waking up to an alarm. In one month, she woke up to her alarm only 9 of the 27 mornings that it went off. What percent of mornings did Sheila wake up to her alarm?
6. The Tigers won 18 of 45 football games this season. What percent of the games did the Tigers win?
7. Twenty percent of Ms. Nakata's class is left-handed. If there are five left-handed students in Ms. Nakata's class, how many students are there in the class?
8. Special Agent Tate earns $26,000 a year. If she receives a 4.5% salary increase, how much will she earn?
9. In a given area of the United States, in one year, there were about 215 highway accidents associated with drinking alcohol. Of these, 113 were caused by speeding. About what percent of the alcohol-related accidents were caused by speeding?
10. A merchant buys a product for $12.20 and then marks it up 35% to sell it. What is the selling price of the item?
11. Natasha bought a set of golf clubs for $340. If she sold them for $255, what was her percent loss?
12. A dealer buys a car from the manufacturer for $13,000. If the dealer wants to earn a profit of 20% based on the cost, at what price should he sell the car?

SECTION

4

Answers & Explanations

LESSON 9

1. First, drop the percent sign to get 25.
   Then, move the decimal point two places to the left. Your final answer is 0.25.
2. First, drop the percent sign: 12.
   Then, move the decimal point two places to the left. Your final answer is 0.12.
3. First, drop the percent sign: 50.
   Then, move the decimal point two places to the left. Your final answer is 0.50 or 0.5.
4. First, drop the percent sign: 62.5.
   Then, move the decimal point two places to the left. Your final answer is 0.625.
5. First, drop the percent sign: 4.2.
   Then, move the decimal point two places to the left. Your final answer is 0.042.
6. First, drop the percent sign: 0.2.
   Then, move the decimal point two places to the left. Your final answer is 0.002.
7. First, drop the percent sign: 125.
   Then, move the decimal point two places to the left. Your final answer is 1.25.

8. First, drop the percent sign: 100.
Then, move the decimal point two places to the left. Your final answer is 1.00 or 1.

9. First, drop the percent sign: 128.9.
Then, move the decimal point two places to the left. Your final answer is 1.289.

10. First, drop the percent sign: 2,000.
Then, move the decimal point two places to the left. Your final answer is 20.00 or 20.

11. First, move the decimal point two places to the right: 040.
Then, delete the zero in front of the 4 and add a percent sign after the number: 40%.

12. First, move the decimal point two places to the right: 075.
Then, delete the zero in front of the number and add a percent sign after it: 75%.

13. First, move the decimal point two places to the right: 062.5.
Then, delete the zero in front of the number and add a percent sign after it: 62.5%.

14. First, move the decimal point two places to the right: 029.
Then, delete the zero in front of the number and add a percent sign after it: 29%.

15. First, move the decimal point two places to the right: 033.
Then, delete the zero in front of the number and add a percent sign after it: 33%.

16. First, move the decimal point two places to the right: 156.
Then, add a percent sign after the number: 156%.

17. First, move the decimal point two places to the right (you'll have to add a zero to the end of the number first): 200.
Then, add a percent sign after the number: 200%.

18. First, move the decimal point two places to the right (you'll have to add a zero to the end of the number first): 650.
Then, add a percent sign after the number: 650%.

19. First, move the decimal point two places to the right: 356.
Then, add a percent sign after the number: 356%.

20. Start by putting the decimal point in the number and adding two zeros: 8.00.
Then, move the decimal point two places to the right: 800. Finally, add a percent sign at the end. Your final answer is 800%.

21. First, drop the percent sign, and write the number as a fraction over 100: $\frac{16}{100}$. Then, reduce the fraction to lowest terms. Your final answer is $\frac{4}{25}$.

22. First, drop the percent sign, and write the number as a fraction over 100: $\frac{5}{100}$. Then, reduce the fraction to lowest terms: $\frac{5}{100} = \frac{1}{20}$. Your final answer is $\frac{1}{20}$.

23. First, drop the percent sign, and write the number as a fraction over 100: $\frac{25}{100}$. Then, reduce the fraction to lowest terms: $\frac{25}{100} = \frac{1}{4}$. Your final answer is $\frac{1}{4}$.

24. First, drop the percent sign, and write the number as a fraction over 100: $\frac{80}{100}$. Then, reduce the fraction to lowest terms: $\frac{80}{100} = \frac{8}{10} = \frac{4}{5}$. Your final answer is $\frac{4}{5}$.

25. First, drop the percent sign, and write the number as a fraction over 100: $\frac{34}{100}$. Then, reduce the fraction to lowest terms: $\frac{34}{100} = \frac{17}{50}$. Your final answer is $\frac{17}{50}$.

26. First, drop the percent sign, and write the number as a fraction over 100: $\frac{10}{100}$. Then, reduce the fraction to lowest terms: $\frac{10}{100} = \frac{1}{10}$. Your final answer is $\frac{1}{10}$.

27. Drop the percent sign, and write the number as a fraction over 100: $\frac{89}{100}$. Since the fraction is already in lowest terms, your final answer is $\frac{89}{100}$.

28. Drop the percent sign, and write the number as a fraction over 100: $\frac{3}{100}$. Since the fraction is already in lowest terms, your final answer is $\frac{3}{100}$.

29. First, drop the percent sign, and write the number as a fraction over 100: $\frac{87\frac{1}{2}}{100}$. Then, rewrite the fraction as a division problem: $87\frac{1}{2} \div 100$. Next, change the mixed number $87\frac{1}{2}$ to an improper fraction. $\frac{175}{2} \div 100$. Invert the second fraction and multiply: $\frac{175}{2} \times \frac{1}{100} = \frac{175}{200}$. Finally, reduce the fraction to lowest terms. Your final answer is $\frac{7}{8}$.

30. First, drop the percent sign, and write the number as a fraction over 100: $\frac{16\frac{2}{3}}{100}$. Then, rewrite the fraction as a division problem: $16\frac{2}{3} \div 100$. Next, change the mixed number $16\frac{2}{3}$ to an improper fraction. $\frac{50}{3} \div 100$. Invert the second fraction and multiply: $\frac{50}{3} \times \frac{1}{100} = \frac{50}{300}$. Finally, reduce the fraction to lowest terms. Your final answer is $\frac{1}{6}$.

31. You know that $\frac{3}{4}$ is equal to 0.75. Then, you just have to move the decimal point to the right and add a percent sign. So your final answer is 75%.

32. You know that $\frac{1}{2}$ is equal to 0.50. Then, you just have to move the decimal point to the right and add a percent sign. So your final answer is 50%.

33. If you don't already know that $\frac{3}{5}$ is equal to 0.60, then use either Method 1 or Method 2 to convert $\frac{3}{5}$ to a percent. If you use Method 1, you first multiply the fraction by $\frac{100}{1}$:
$\frac{3}{5} \times \frac{100}{1} = \frac{300}{5} = \frac{60}{1}$
You know that $\frac{60}{1} = 60$. Finally, add a percent sign, so your final answer is 60%.

34. You know that $\frac{1}{4}$ is equal to 0.25. Then, you just have to move the decimal point to the right and add a percent sign. So your final answer is 25%.

35. One way to solve this problem is to first multiply the fraction by $\frac{100}{1}$:
$\frac{300}{50} \times \frac{100}{1} = \frac{300}{50} = \frac{6}{1}$
You know that $\frac{6}{1} = 6$. Finally, add a percent sign, so your final answer is 6%.

36. One way to solve this problem is to first divide the numerator by the denominator. If you divide 1 by 8, you get 0.125. Then, move the decimal point two places to the right (this is the same as multiplying by 100) and add a percent sign. So your final answer is 12.5%.

37. One way to solve this problem is to first multiply the fraction by $\frac{100}{1}$:
$\frac{7}{10} \times \frac{100}{1} = \frac{700}{10} = \frac{70}{1}$
You know that $\frac{70}{1} = 70$.
Finally, add a percent sign, so your final answer is 70%.

38. One way to solve this problem is to first multiply the fraction by $\frac{100}{1}$:
$\frac{17}{20} \times \frac{100}{1} = \frac{1700}{20} = \frac{85}{1}$
You know that $\frac{85}{1} = 85$. Finally, add a percent sign, so your final answer is 85%.

39. One way to solve this problem is to first multiply the fraction by $\frac{100}{1}$:
$\frac{19}{25} \times \frac{100}{1} = \frac{1900}{25} = \frac{76}{1}$
You know that $\frac{76}{1}$ = 76. Finally, add a percent sign, so your final answer is 76%.

40. One way to solve this problem is to first multiply the fraction by $\frac{100}{1}$:
$\frac{18}{5} \times \frac{100}{1} = \frac{1800}{5} = \frac{360}{1}$
You know that $\frac{360}{1}$ = 360. Finally, add a percent sign, so your final answer is 360%.

41. Begin by changing the mixed number to an improper fraction: $3\frac{1}{4}$ becomes $\frac{13}{4}$. Then, multiply the fraction by $\frac{100}{1}$.
$\frac{13}{4} \times \frac{100}{1} = \frac{1300}{4} = \frac{320}{1}$
You know that $\frac{325}{1}$ = 325, so you add a percent sign and your final answer is 325%.

42. Begin by changing the mixed number to an improper fraction: $9\frac{4}{5}$ becomes $\frac{49}{5}$. Then, multiply the fraction by $\frac{100}{1}$.
$\frac{49}{5} \times \frac{100}{1} = \frac{4900}{5} = \frac{980}{1}$
You know that $\frac{980}{1}$ = 980, so you add a percent sign and your final answer is 980%.

LESSON 10

1. Begin by figuring out what you know from the problem and what you're looking for.
- You have the percent: 1%.
- You have the whole: 34.
- You are looking for the part.

Then, use the equation to solve the problem: Whole × Percent = Part. Plug in the pieces of the equation that you know: Part = 34 × 0.01 = 0.34. The answer is 0.34.

2. Begin by figuring out what you know from the problem and what you're looking for.
- You have the percent: 10%.
- You have the whole: 52.
- You are looking for the part.

Then, use the equation to solve the problem: Whole × Percent = Part. Plug in the pieces of the equation that you know: Part = 52 × 0.1 = 5.2. The answer is 5.2.

3. Begin by figuring out what you know from the problem and what you're looking for.
- You have the percent: 0.5%.
- You have the whole: 30.
- You are looking for the part.

Then, use the equation to solve the problem: Whole × Percent = Part. Plug in the pieces of the equation that you know: Part = 30 × 0.005 = 0.15. The answer is 0.15.

4. Begin by figuring out what you know from the problem and what you're looking for.
- ▶ You have the percent: 100%.
- ▶ You have the whole: 99.
- ▶ You are looking for the part.

Then, use the equation to solve the problem: Whole × Percent = Part. Plug in the pieces of the equation that you know: Part = 99 × 1.00 = 99. The answer is 99.

5. Begin by figuring out what you know from the problem and what you're looking for.
- ▶ You have the percent: 25%.
- ▶ You have the whole: 100.
- ▶ You are looking for the part.

Then, use the equation to solve the problem: Whole × Percent = Part. Plug in the pieces of the equation that you know: Part = 100 × 0.25 = 25. The answer is 25.

Remember when multiplying by a multiple of 10, you just have to move the decimal point the correct number of places.

6. Begin by figuring out what you know from the problem and what you're looking for.
- ▶ You have the percent: 20%.
- ▶ You have the whole: 70.
- ▶ You are looking for the part.

Then, use the equation to solve the problem: Whole × Percent = Part. Plug in the pieces of the equation that you know: 70 × 0.20 = Part = 14. The answer is 14.

7. Begin by figuring out what you know from the problem and what you're looking for.
- ▶ You have the percent: 90%.
- ▶ You have the whole: 10.
- ▶ You are looking for the part.

Then, use the equation to solve the problem: Whole × Percent = Part. Plug in the pieces of the equation that you know: 10 × 0.90 = Part = 9. The answer is 9.

Remember when multiplying by a multiple of 10, you just have to move the decimal point.

8. Begin by figuring out what you know from the problem and what you're looking for.
- ▶ You have the percent: 80%.
- ▶ You have the whole: 50.
- ▶ You are looking for the part.

Then, use the equation to solve the problem: Whole × Percent = Part. Plug in the pieces of the equation that you know: 50 × 0.80 = Part = 40. The answer is 40.

9. Begin by figuring out what you know from the problem and what you're looking for.
- ▶ You have the percent: 50%.
- ▶ You have the whole: 120.
- ▶ You are looking for the part.

Then, use the equation to solve the problem: Whole × Percent = Part. Plug in the pieces of the equation that you know: 120 × 0.50 = Part = 60. Thus, 60 students received a B last year in Mr. Hopkin's class.

10. Begin by figuring out what you know from the problem and what you're looking for.
- ▶ You have the percent: 40%.
- ▶ You have the whole: 500.
- ▶ You are looking for the part.

Then, use the equation to solve the problem: Whole × Percent = Part. Plug in the pieces of the equation that you know: 500 × 0.40 = Part = 200. The answer is 200 candy bars.

11. Begin by figuring out what you know from the problem and what you're looking for.
- ▶ You have the part: 10.
- ▶ You have the whole: 50.
- ▶ Since you are looking for the percent, write a fraction of the part over the whole: $\frac{10}{50}$.

Then, convert the fraction to a percent:
10 ÷ 50 = 0.2
0.2 × 100 = 20

The answer is 20%.

12. Begin by figuring out what you know from the problem and what you're looking for.
- ▶ You have the part: 20.
- ▶ You have the whole: 80.
- ▶ Since you are looking for the percent, write a fraction of the part over the whole: $\frac{20}{80}$.

Then, convert the fraction to a percent:
$20 \div 80 = 0.25$
$0.25 \times 100 = 25$

The answer is 25%.

13. Begin by figuring out what you know from the problem and what you're looking for.
- ▶ You have the part: 5.
- ▶ You have the whole: 20.
- ▶ Since you are looking for the percent, write a fraction of the part over the whole: $\frac{5}{20}$.

Then, convert the fraction to a percent:
$5 \div 20 = 0.25$
$0.25 \times 100 = 25$

The answer is 25%.

14. Begin by figuring out what you know from the problem and what you're looking for.
- ▶ You have the part: 6.
- ▶ You have the whole: 6.
- ▶ Since you are looking for the percent, write a fraction of the part over the whole: $\frac{6}{6}$.

Then, convert the fraction to a percent:
$6 \div 6 = 1$
$1 \times 100 = 100$

The answer is 100%.

15. Begin by figuring out what you know from the problem and what you're looking for.
- ▶ You have the part: 9.
- ▶ You have the whole: 18.
- ▶ Since you are looking for the percent, write a fraction of the part over the whole: $\frac{9}{18}$.

Then, convert the fraction to a percent:
$9 \div 18 = 0.5$
$0.5 \times 100 = 50$

The answer is 50%.

16. Begin by figuring out what you know from the problem and what you're looking for.

▶ You have the part: 1.
▶ You have the whole: 50.
▶ Since you are looking for the percent, write a fraction of the part over the whole: $\frac{1}{50}$.

Then, convert the fraction to a percent:

$1 \div 50 = 0.02$
$0.02 \times 100 = 2$

The answer is 2%.

17. Begin by figuring out what you know from the problem and what you're looking for.

▶ You have the part: 22.
▶ You have the whole: 100.
▶ Since you are looking for the percent, write a fraction of the part over the whole: $\frac{22}{100}$.

Then, convert the fraction to a percent:
$22 \div 100 = 0.22$
$0.22 \times 100 = 22$

The answer is 22%.

18. Begin by figuring out what you know from the problem and what you're looking for.

▶ You have the part: 3.
▶ You have the whole: 30.
▶ Since you are looking for the percent, write a fraction of the part over the whole: $\frac{3}{30}$.

Then, convert the fraction to a percent:
$3 \div 30 = 0.1$
$0.1 \times 100 = 10$

The answer is 10% of the class.

19. Begin by figuring out what you know from the problem and what you're looking for.

▶ You have the part: 12.
▶ You have the whole: 60.
▶ Since you are looking for the percent, write a fraction of the part over the whole: $\frac{12}{60}$.

Then, convert the fraction to a percent:
$12 \div 60 = 0.2$
$0.2 \times 100 = 20$

Thus, 20% of those trying out are selected.

20. Begin by figuring out what you know from the problem and what you're looking for.

- You have the part: 5 (100 – 95 = 5).
- You have the whole: 100.
- Since you are looking for the percent, write a fraction of the part over the whole: $\frac{5}{100}$.

You know that $\frac{5}{100}$ = 5%, so the answer is 5%.

21. Begin by figuring out what you know from the problem and what you're looking for.

- You have the part: 20.
- You also have the percent: 20%.
- Since you are looking for the whole, write the part over the percent: $\frac{20}{20\%}$.

Then, convert the percent to fractions to make your division easier: $\frac{20}{1} \div \frac{20}{100}$ = Whole. Invert the second fraction and multiply: $\frac{20}{1} \times \frac{100}{20}$. You can cancel the 20s, and then multiply: $\frac{\overset{1}{\cancel{20}}}{1} \times \frac{100}{\underset{1}{\cancel{20}}}$. Now that your problem is simplified, multiply across the top and bottom to get your final answer: 100.

22. Begin by figuring out what you know from the problem and what you're looking for.

- You have the part: 30.
- You also have the percent: 75%.
- Since you are looking for the whole, write the part over the percent:

$\frac{30}{75\%}$.

Then, convert the percent to a fraction to make your division easier: 30 ÷ $\frac{75}{100}$ = Whole. Invert the second fraction and multiply: $\frac{30}{1} \times \frac{100}{75}$. Multiply across the top and bottom to get your final answer. So, the answer is 40.

23. Begin by figuring out what you know from the problem and what you're looking for.

- You have the part: 6.
- You also have the percent: 50%.
- Since you are looking for the whole, write the part over the percent:

$\frac{6}{50\%}$.

Then, convert the percent to a fraction to make your division easier: 6 ÷ $\frac{50}{100}$ = Whole. Invert the second fraction and multiply: $\frac{6}{1} \times \frac{100}{50}$. Multiply across the top and bottom to get your final answer: 12.

24. Begin by figuring out what you know from the problem and what you're looking for.

- ▶ You have the part: 75.
- ▶ You also have the percent: 75%.
- ▶ Since you are looking for the whole, write the part over the percent:

$\frac{75}{75\%}$.

Then, convert the percent to a fraction to make your division easier: $75 \div \frac{75}{100}$ = Whole

Invert the second fraction and multiply: $\frac{75}{1} \times \frac{100}{75}$. You can cancel the 75s, and then multiply: $\frac{\cancel{75}^{1}}{1} \times \frac{100}{\cancel{75}_{1}}$. Now that your problem is simplified, multiply across the top and bottom to get your final answer: 100.

25. Begin by figuring out what you know from the problem and what you're looking for.

- ▶ You have the part: 20.
- ▶ You also have the percent: 2%.
- ▶ Since you are looking for the whole, write the part over the percent:

$\frac{20}{2\%}$.

Then, convert the percent to a fraction to make your division easier: $20 \div \frac{2}{100}$ = Whole.

Invert the second fraction and multiply: $\frac{20}{1} \times \frac{100}{2}$. Multiply across the top and bottom to get your final answer: 1,000.

26. Begin by figuring out what you know from the problem and what you're looking for.

- ▶ You have the part: 100.
- ▶ You also have the percent: 40%.
- ▶ Since you are looking for the whole, write the part over the percent:

$\frac{100}{40\%}$.

Then, convert the percent to a fraction to make your division easier: $100 \div \frac{40}{100}$ = Whole. Invert the second fraction and multiply: $\frac{100}{1} \times \frac{100}{40}$. You can cancel to simplify the multiplication: $\frac{100}{1} \times \frac{100}{40}$. Multiply across the top and bottom to get your final answer: 250.

27. Begin by figuring out what you know from the problem and what you're looking for.

- ▶ You have the part: 5.
- ▶ You also have the percent: 10%.
- ▶ Since you are looking for the whole, write the part over the percent:

$$\frac{5}{10\%}.$$

Then, convert the percent to a fraction to make your division easier: $5 \div \frac{10}{100}$ = Whole. Invert the second fraction and multiply: $\frac{5}{1} \times \frac{100}{10}$. Multiply across the top and bottom to get your final answer: 50.

28. Begin by figuring out what you know from the problem and what you're looking for.

- ▶ You have the part: 120.
- ▶ You also have the percent: 80%.
- ▶ Since you are looking for the whole, write the part over the percent:

$$\frac{120}{80\%}.$$

Then, convert the percent to a fraction to make your division easier: $120 \div \frac{80}{100}$ = Whole.

Then, convert the percent to a fraction to make your division easier: $120 \div \frac{80}{100}$ = Whole. Invert the second fraction and multiply: $\frac{120}{1} \times \frac{100}{80}$. You can cancel to simplify the multiplication: $\frac{\overset{6}{\cancel{120}}}{1} \times \frac{100}{\underset{4}{\cancel{80}}}$. Multiply across the top and bottom to get your final answer: 150.

29. Begin by figuring out what you know from the problem and what you're looking for.

- ▶ You have the part: 50.
- ▶ You also have the percent: 200%.
- ▶ Since you are looking for the whole, write the part over the percent:

$$\frac{50}{200\%}.$$

Then, convert the percent to a fraction to make your division easier: $50 \div \frac{200}{100}$ = Whole.

Invert the second fraction and multiply: $\frac{50}{1} \times \frac{100}{200}$. You can cancel to simplify the multiplication: $\frac{\overset{1}{\cancel{50}}}{1} \times \frac{100}{\underset{4}{\cancel{200}}}$. Multiply across the top and bottom to get your final answer: 25.

30. Begin by figuring out what you know from the problem and what you're looking for.

- ▶ You have the part: 1.
- ▶ You also have the percent: 2%.
- ▶ Since you are looking for the whole, write the part over the percent:

$$\frac{1}{2\%}.$$

Then, convert the percent to a fraction to make your division easier: $1 \div \frac{2}{100}$ = Whole.

Invert the second fraction and multiply: $\frac{1}{1} \times \frac{100}{2}$. Multiply across the top and bottom to get 50. So, the answer is 50 students.

REAL WORLD PROBLEMS

1. Begin by figuring out what you know from the problem and what you're looking for.
- ▶ You have the percent: 25%.
- ▶ You have the whole: $50.
- ▶ You are looking for the part.

Part = $50 × 0.25 = $12.50. The answer is $12.50.

2. Begin by figuring out what you know from the problem and what you're looking for.
- ▶ You have the percent: 90%.
- ▶ You have the whole: $800.
- ▶ You are looking for the part.

Part = $800 × 0.90 = $720—this is the part that the insurance company will pay. The question asks for the part that Estrella has to pay: $800 – ($800 × 0.90) = $80. The answer is $80.

Another way to look at the problem: If the insurance company will pay 90%, then Estrella has to pay 10% (100% – 90% = 10%). Estrella has to pay $80 ($800 × 0.10 = $80).

3. Begin by figuring out what you know from the problem and what you're looking for.
- ▶ You have the percent: 33%.
- ▶ You have the whole: 20 ounces.
- ▶ You are looking for the part.

Part = 20 ounces × 0.33 = 6.6 ounces. The answer is 6.6 ounces.

4. Begin by figuring out what you know from the problem and what you're looking for.
- ▶ You have the part: 56.
- ▶ You also have the percent: 80%.
- ▶ Since you are looking for the whole (total number of items on the test), write the part over the percent: $\frac{56}{0.80}$ and divide.

So, there were 70 items on the test.

5. Begin by figuring out what you know from the problem and what you're looking for.

- ▶ You have the part: 9.
- ▶ You have the whole: 27.
- ▶ Since you are looking for the percent, write a fraction of the part over the whole and convert it to a percent: $\frac{9}{27} \times 100 = 33\frac{1}{3}\%$

Thus, Sheila woke up to her alarm only $33\frac{1}{3}$ % of the mornings.

6. Begin by figuring out what you know from the problem and what you're looking for.

- ▶ You have the part: 18.
- ▶ You have the whole: 45.
- ▶ Since you are looking for the percent, write a fraction of the part over the whole and convert it to a percent: $\frac{18}{45} \times 100 = 40\%$.

Thus, the Tigers won 40% of the football games.

7. Begin by figuring out what you know from the problem and what you're looking for.

- ▶ You have the part: 5.
- ▶ You also have the percent: 20%.
- ▶ Since you are looking for the whole, write the part over the percent ($\frac{5}{0.20}$) and divide.

So, there are 25 students in the class.

8. Begin by figuring out what you know from the problem and what you're looking for.

- ▶ You have the percent: 4.5%.
- ▶ You have the whole: $26,000.
- ▶ You are looking for the part: Part = $26,000 × 0.045 = $1,170.

But the question asks for the amount she will make with this raise, so you have to add the part to her salary: $1,170 + $26,000 = $27,170.

So she will earn $27,170.

Another way to look at the problem: Special Agent Tate will earn $26,000 × 0.045 + $26,000 = $27,170.

9. Begin by figuring out what you know from the problem and what you're looking for.

- You have the whole: 215.
- Since you are looking for the percent, write a fraction of the part over the whole and convert it to a percent: $\frac{113}{215} \times 100 = 52.56\%$, or about 53%.

Thus, about 53% of alcohol-related accidents are also caused by speeding.

10. Begin by figuring out what you know from the problem and what you're looking for.

- You have the percent: 35%.
- You have the whole: $12.20.
- You are looking for the part: Part = $12.20 × 0.35 = $4.27. This is the amount of mark-up the merchant will add to the price of the product.

You have to add the part to the price: $4.27 + $12.20 = $16.47. So the selling price will be $16.47.

Another way to look at the problem: The selling price is $16.47 ($12.20 × 0.035 + $12.20).

11. Begin by figuring out what you know from the problem and what you're looking for.

- You have the part: $340 – $255 = $85.
- You have the whole: $340.
- Since you are looking for the percent, write a fraction of the part over the whole and convert it to a percent: $\frac{85}{340} \times 100 = 25\%$.

So, she lost 25%.

Another way to look at the problem:

$$\frac{\$340 - \$255}{\$340} \times 100 = 25\%$$

12. Begin by figuring out what you know from the problem and what you're looking for.

- You have the percent: 20%.
- You have the whole: $13,000.
- You are looking for the part: Part = $13,000 × 0.20 = $2,600. This is the amount of profit the car dealer wants to make.

You have to add the part to the price: $2,600 + $13,000 = $15,600. So the selling price will be $15,600.

Another way to look at the problem: He should sell the car for $13,000 × 0.20 + $13,000 = $15,600.

SECTION

5

Statistics

STATISTICS ARE EVERYWHERE—in news reports, sports, and on your favorite websites. Mean, median, and mode are three common statistics that give information on a group of numbers. They are called *measures of central tendency* because they are different ways of finding the central trend in a group of numbers. Ratios and proportions are ways to compare these statistics.

Similarly, you see probabilities or predictions all the time. Listening to the weather report, you may hear that there is a 60% chance of rain tomorrow. At karate lessons, you may hear that 19 out of 20 advanced students will attain a brown belt. On television, you might hear that four out of five dentists recommend a certain toothbrush. These are all ways to express probability. In this section, you will also learn what probability is and how to calculate it.

LESSON

11

Finding Mean, Median, and Mode

LESSON SUMMARY

In this lesson, you will learn how to find the mean, median, and mode of a set of numbers.

Statistics are everywhere. In news reports, at the doctor's office, even in class at school—we get information in the form of statistics every day. Mean, median, and mode are three common statistics that give information on of a group of numbers. In fact, they are called *measures of central tendency* because they are different ways of finding the central trend of a group of numbers.

FINDING THE MEAN

Mean is just another word for *average*. The mean, or average, is one of the most useful and common statistics. You probably already average your grades at school regularly, so you may already know the basic steps to finding the mean of a set of numbers.

Step 1: Add all the numbers in the list.

Step 2: Count the number of numbers in the list.

Step 3: Divide the sum (the result of Step 1) by the number (the result of Step 2).

Another way to think about the mean is in the form of this equation:

$$Mean = \frac{\textit{the sum of the numbers}}{\textit{the number of numbers}}$$

Example: Find the mean of the following set of numbers: 5, 7, 19, 12, 4, 11, 15.

Step 1: Add all the numbers in the list.

5 + 7 + 19 + 12 + 4 + 11 + 15 = 73

Step 2: Count the number of numbers in the list.

There are seven numbers in the list.

Step 3: Divide the sum (the result of Step 1) by the number (the result of Step 2).

$\frac{73}{7} = 10.4$

So, the mean is 10.4.

Example: Jason has four grades of equal weight in history. They are 82, 90, 88, and 85. What is Jason's mean (average) in history?

Step 1: Add all the numbers in the list.

82 + 90 + 88 + 85 = 345

Step 2: Count the number of numbers in the list.

There are four numbers in the list.

Step 3: Divide the sum (the result of Step 1) by the number (the result of Step 2).

$\frac{345}{4} = 86.25$

So, the mean is 86.25.

PRACTICE

Find the mean for each set of numbers. You can check your answers at the end of the section.

1. 3, 7, 8, 10, 3, 5, 6

2. 23, 45, 67, 48, 36

3. 88, 92, 100, 95, 85

4. 3, 4, 5, 6, 7, 8, 9, 6

5. 100, 105, 110, 101, 103, 107

Solve each problem. You can check your answers at the end of the section.

6. Misha has ten grades of equal weight in English. They are 82, 85, 72, 90, 88, 86, 91, 93, 81, and 85. What is Misha's mean (average) grade in English?

7. So far this grading period, Jack has three grades of equal weight in Math. They are 82, 81, and 84. He needs an 83 average or better to receive a B in Math. Does he have a B in Math so far?

8. Mara babysat all four weekends last month. She earned the following amounts each weekend:
First weekend: $25
Second weekend: $27
Third weekend: $13
Fourth weekend: $18

What is the mean of Mara's earnings per weekend last month?

9. John spent the following amounts on his brothers' birthday presents this year:
Mike: $21
Dave: $17
Steve: $19

On average, how much did John spend per brother on birthday presents this year?

10. Philip saves a portion of his allowance each week. Last month, he saved $5, $7, $2, and $9. On average, how much did Philip save per week last month?

▶ FINDING THE MEDIAN

The *median* is the middle number in a group of numbers arranged in sequential order. In a set of numbers, half will be greater than the median and half will be less than the median.

Step 1: Put the numbers in sequential order.
Step 2: The middle number is the median.

Example: Find the median of the following set of numbers: 5, 7, 19, 12, 4, 11, 15.
Step 1: Put the numbers in sequential order.
4, 5, 7, 11, 12, 15, 19
Step 2: The middle number is the median.
The middle number is 11.
So, 11 is the median.

In the last example, there was an odd number of numbers, so the middle number was easy to find. But what if you are given an even number of numbers? Let's see how it works.

EXAMPLE: Find the median of the following set of numbers: 5, 7, 19, 12, 4, 11, 15, 13.

Step 1: Put the numbers in sequential order.

4, 5, 7, 11, 12, 13, 15, 19

Step 2: The middle number is the median.

But there are two middle numbers: 11 and 12. In this case, you find the *mean* (or average) of the two middle numbers. That value is your *median*. Remember, to find the mean of a set of numbers, you first add the numbers together (11 + 12 = 23). Then, you divide the sum by the number of numbers (23 ÷ 2 = 11.5).

So, 11.5 is the median.

PRACTICE

Find the median for each set of numbers. You can check your answers at the end of the section.

11. 3, 7, 8, 10, 3, 5, 6

12. 23, 45, 67, 48, 36

13. 88, 92, 100, 95, 85

14. 3, 4, 5, 6, 7, 8, 9, 10, 11

15. 100, 105, 104, 110, 101, 103, 107

16. 12, 17, 11, 14, 16

17. 3, 7, 4, 3, 8, 9

18. 7, 8, 5, 6, 9, 10

Solve each problem. You can check your answers at the end of the section.

19. Jill scored the following points in her last five volleyball games: 3, 5, 2, 6, 1. What is the median of her point totals?

20. Jack has three grades in Math. They are 82, 81, and 84. What is the median of his grades in Math?

Why would you use the median instead of the mean? Let's say your teacher gives everyone above the class mean either an A or B. Here are the grades on the last test.

GRADES		
110	70	65
80	70	64
79	69	63
78	68	60
75	67	52
72	65	

The class mean is 71, so only six students will receive an A or a B on the last test. All the other students will get a C or below. How would the result be different if the teacher used the class median to determine who gets an A and who gets a B? In that case, everyone with a test score greater than a 69 would get either an A or a B on the test—that's eight students. About half the students would get an A or a B using the median.

Notice that the mean was raised by the one person who received a 110 on the test. Often, when one number changes the mean to be higher than the center value, the median can be used instead.

FINDING THE MODE

The mode refers to the number in a set of numbers that occurs most frequently. To find the mode, you just look for numbers that occur more than once and find the one that appears *most* often.

Example: Find the mode of the following set of numbers: 5, 7, 9, 12, 9, 11, 15.
The number 9 occurs twice in the list, so 9 is the mode.

Example: Find the mode of the following set of numbers: 5, 7, 19, 12, 4, 11, 15.
None of the numbers occurs more than once, so there is no mode.

Example: Find the mode of the following set of numbers: 5, 7, 9, 12, 9, 11, 5.
The numbers 5 and 9 both occur twice in the list, so both 5 and 9 are modes. When a set of numbers has two modes, it is called *bimodal.*

As you can see, the mode isn't always a middle number in a set of numbers. Instead, mode shows clustering. Mode is often used in stores to decide which sizes, styles, or prices are most popular. It wouldn't make sense for a clothing store to stock up on the mean size or the median size of pants. It makes more sense to buy the sizes that most people wear. And how would a store find the mean or the median of different styles of clothes? It can't. There's where the mode comes in.

PRACTICE

Find the mode for each set of numbers. You can check your answers at the end of the section.

21. 3, 7, 8, 10, 3, 5, 6

22. 23, 45, 67, 48, 36, 45

23. 2, 5, 7, 8, 3, 5

24. 3, 4, 5, 6, 7, 8, 9, 10, 11

25. 100, 103, 104, 110, 101, 103, 107

26. 12, 17, 11, 14, 13, 12, 17, 16

27. 3, 7, 4, 3, 8, 9, 7

28. 7, 8, 5, 6, 9, 10

29. 25, 27, 39, 22, 39, 11, 25

30. 3, 7, 4, 3, 8, 9, 7, 7, 8, 5, 6, 9, 10, 3, 8

Using Ratios and Proportion

LESSON SUMMARY

In this lesson, you will learn what ratios and proportions are. You will also learn how to write ratios and proportions and how to work with them in basic calculations, including those used in measurement conversions.

Ratios and proportions are often found in textbooks and news reports. You'll also find them in math word problems.

WHAT ARE RATIOS?

A *ratio* is a way of comparing two or more numbers. There are several different ways to write ratios. Here are some examples of ways to write ratios.

- with the word *to*: 1 to 2
- using a colon (:) to separate the numbers: 1 : 2
- using the term *for every*: 1 for every 2
- separated by a division sign or fraction bar: $\frac{1}{2}$

Example: Write the following ratio as a fraction: five girls to six boys.
The question asks you to write the ratio as a fraction: $\frac{5}{6}$.

Keep the terms of a ratio in the order that the problem compares them. Since the last example compared girls to boys, your answer should give the number of girls to the number of boys.

Example: Write the following ratio using a colon: 10 wins to 5 losses.

10:5

Example: Write the following ratio as a fraction: 10 wins to 5 losses.

$\frac{10}{5}$

Ratios often look like fractions and they can be reduced to lowest terms just as fractions can be. However, do not change a ratio whose numerator is larger than its denominator to a whole number or a mixed number. A ratio that looks like a fraction is comparing the two numbers. You will lose the comparison if you treat the ratio as a fraction and convert it to a mixed number. Let's look again at the last example.

EXAMPLE: Write the following ratio as a fraction: 10 wins to 5 losses.

$\frac{10}{5}$

Even though $\frac{10}{5}$ looks like an improper fraction, it's not here—it's a ratio comparing the number of wins to the number of losses. You can however, reduce the ratio to lowest terms: $\frac{10}{5} = \frac{2}{1}$.

PRACTICE

Write the following ratios as fractions. You can check your answers at the end of the section.

1. 12 wins to 15 losses

2. 15 girls to 20 boys

3. 5 umbrellas for 10 people

4. 6 red beads to 4 yellow beads

5. two cups of sugar for every batch of cookies

Write the following ratios using a colon. You can check your answers at the end of the section.

6. 5 teachers to 40 students

7. one head to one tail

8. three people failed for every nine who passed

9. three correct answers for every incorrect answer

10. two cat's eye marbles to 10 blue marbles

SOLVING RATIO PROBLEMS

There are several kinds of ratio problems. The examples below show how to solve different kinds of ratio problems.

Example: A painter mixes two quarts of red paint to three quarts of white paint. What is the ratio of red paint to white paint?

There are several ways you could write this ratio:

2 quarts of red paint to 3 quarts of white paint, or 2 to 3

2 quarts red paint: 3 quarts white paint, or 2:3

$\frac{\text{2 quarts red paint}}{\text{3 quarts white paint}}$, or $\frac{2}{3}$

Example: Last season, the Tigers won 30 games. They lost only 6 games. There were no tied games last season.

a. What is the ratio of games won to games lost?

b. What is the ratio of games won to games played?

Write your answers as fractions.

a. The first part of the question asks for the ratio of games won to games lost. So, you would write $\frac{30}{6}$. You could reduce the ratio to $\frac{5}{1}$.

b. The second part of the question asks for the ratio of games won to games played. First, you need to calculate the total number of games played. Since the Tigers won 30 games, lost 6 games and tied no games, they must have played a total of 36 games. The ratio of games won to games played is

$\frac{\text{30 games won}}{\text{36 total games}}$, or $\frac{30}{36}$

You could reduce $\frac{30}{36}$ to $\frac{5}{6}$.

Often, ratios are written to look like fractions. For example, 2:3 can be written as $\frac{2}{3}$; 7 to 8 can be written as $\frac{7}{8}$; and so on. In Section 2, you learned that fractions are ways of representing a part of a whole. Usually, a fraction represents a part over a whole:

$\frac{\text{Part}}{\text{Whole}}$

Often, a ratio represents a part over a part:

$\frac{\text{Part}}{\text{Part}}$

But ratios can also represent a part over a whole:

$\frac{\text{Part}}{\text{Whole}}$

When a ratio represents a part over a part, you can often find the whole if you know all the parts. In the last example, you did know all the parts, so you were able to calculate the total number of games played last season—the whole. Let's look at another example.

EXAMPLE: Assume that there are 15 boys and 20 girls in your math class. What is the *ratio* of boys to girls in your math class?

You can write the ratio 15:20. You could also write it as a fraction: $\frac{15}{20}$, which reduces to $\frac{3}{4}$. Notice, however, that this fraction represents a part over a part.

continued from previous page

EXAMPLE: Assume that there are 15 boys and 20 girls in your math class. What is the *fraction* of girls in your math class?

To write the fraction of girls in your math class, you have to calculate the total number of students in the class. There are 15 boys and 20 girls in your class, so there are 35 students in your class. That is, the whole is the total number of students in math class (35). There are 20 girls in your math class. So, the fraction of girls in math class is $\frac{20}{35}$, which reduces to $\frac{4}{7}$. This fraction represents a part over a whole.

Example: At the Pumpkin Festival last night there were 4 men, 8 women, and 20 children. What is the ratio of men to women to children? Use colons to write your answer.

The ratio is 4:8:20. You could reduce this ratio (by dividing all the numbers by 4) to 1:2:5. This problem asks you to compare more than two numbers, so a fraction cannot be used.

Example: At the Pumpkin Festival last night there were 4 men, 8 women, and 20 children. What is the ratio of children to total people attending the festival? Use colons to write your answer.

First, calculate the total number of people at the Pumpkin Festival: 4 + 8 + 20 = 32. There were 20 children at the festival, so you would write 20:32. You can reduce this ratio (by dividing each number by 4) to 5:8.

Example: At the Pumpkin Festival last night there were 4 men, 8 women, and 20 children. What is the ratio of women to men? Use colons to write your answer.

There were 8 women and 4 men, so you would write 8:4. You can reduce this ratio (by dividing each number by 4) to 2:1.

Example: At the Pumpkin Festival last night there were 4 men, 8 women, and 20 children. What is the ratio of children to adults? Use colons to write your answer.

First, calculate the number of adults present. There were 8 women and 4 men, so there were 12 adults. There were 20 children, so you would write 20:12. You can reduce this ratio (by dividing by 4) to 5:3.

Look at the answers to the last four example problems. Notice that the numbers are all in different orders and represent different things. In some cases, they represent men, women, children, and the total number of people at the festival. That's why it's often helpful to write the words that go with the numbers. This can help you keep the meaning of each ratio in mind. You may not be able to use the labels in your answer when you are working on a multiple-choice test, but you can use the labels as you work through the problem on your own. Let's quickly go through the last four examples again, this time using the labels in the answers.

EXAMPLE: At the Pumpkin Festival last night there were 4 men, 8 women, and 20 children. What is the ratio of men to women to children? Use colons to write your answer.

The ratio is 4 men:8 women:20 children. You could reduce this ratio (by dividing all the numbers by 4) to 1 man:2 women:5 children. Now it's clear what each number refers to. There was one man to every four women and every five children.

EXAMPLE: At the Pumpkin Festival last night there were 4 men, 8 women, and 20 children. What is the ratio of children to total people attending the festival? Use colons to write your answer.

Fist, calculate the total number of people at the Pumpkin Festival: 4 + 8 + 20 = 32. There were 20 children at the festival, so you would write 20 children:32 total people. You can reduce this ratio (by dividing each number by 4) to 5 children:8 total people. There were five children out of every eight people at the festival.

continued from previous page

EXAMPLE: At the Pumpkin Festival last night there were 4 men, 8 women, and 20 children. What is the ratio of women to men? Use colons to write your answer.

There were 8 women and 4 men, so you would write 8 women:4 men. You can reduce this ratio (by dividing each number by 4) to 2 women:1 man. There were two women for every man at the festival.

EXAMPLE: At the Pumpkin Festival last night there were 4 men, 8 women, and 20 children. What is the ratio of children to adults? Use colons to write your answer.

First, calculate the number of adults present. There were 8 women and 4 men, so there were 12 adults. There were 20 children, so you would write 20 children:12 adults. You can reduce this ratio (by dividing each number by 4) to 5 children:3 adults. There were five children for every three adults at the festival.

In word problems, the terms *per, for every,* and *in (a)* can also indicate that a ratio is being used. For example, you might see that Jacques drove 30 miles *per* gallon of gas. When you see the *per*, you can write the ratio as follows:

$$\frac{\text{30 miles}}{\text{1 gallon}}$$

Let's look at some examples of ratios you might find in a word problem on a test and convert them to fractions.

- **The job pays $10 *per* hour.**

$$\frac{\$10}{\text{1 hour}}$$

continued from previous page

- **The camp provides three meals *per* day.**

$\frac{\text{3 meals}}{\text{1 day}}$

- **Use one cup of sugar *for every* four cups of blueberries.**

$\frac{\text{1 cup sugar}}{\text{4 cups of blueberries}}$

- **There are 1.6 kilometers *in a* mile.**

$\frac{\text{1.6 kilometers}}{\text{1 mile}}$

- **There are 12 inches *per* foot.**

$\frac{\text{12 inches}}{\text{1 foot}}$

- **There are 2.54 centimeters *per* inch.**

$\frac{\text{2.54 centimeters}}{\text{1 inch}}$

Notice that ratios written using the word *per* are almost always fractions over the number 1. These ratios are sometimes called *unit rates* because they are written in terms of one unit.

Example: If a driver drives 100 miles in two hours, what is his average speed?

You can answer this question by setting up a ratio: 100 miles:2 hours. You can reduce this ratio (by dividing by 2) to 50 miles:1 hour. So, the driver drove on average 50 miles per hour.

PRACTICE

Solve the following problems. You can check your answers at the end of the section.

11. On his last science test, Marc answered 21 questions correctly. He missed 4 questions. What is the ratio of correct answers to incorrect answers?

12. On his last science test, Mac answered 21 questions correctly. He missed 4 questions. What is the ratio of correct answers to total questions?

13. If a driver drives 325 miles in five hours, what is his average speed?

14. The State Fair Pet Show included 15 chickens, 5 dogs, 20 hamsters, and 10 cats. What is the ratio of dogs to cats to chickens to hamsters? Write your answer in lowest terms.

15. The State Fair Pet Show included 15 chickens, 5 dogs, 20 hamsters, and 10 cats. What is the ratio of dogs to cats? Write your answer in lowest terms.

16. The State Fair Pet Show included 15 chickens, 5 dogs, 20 hamsters, and 10 cats. What is the ratio of cats to total animals shown? (Assume that only chickens, dogs, cats, and hamsters took part in the pet show.) Write your answer in lowest terms.

WHAT ARE PROPORTIONS?

A *proportion* is a way of relating two ratios to one another. Let's say you read in your school newspaper that 8 out of 10 students at your school are expected to take the PSAT this year. If there are 100 students in your school, then 80 students are expected to take the test this year. This is an example of a proportion. Proportions can be written as equations. For example, this proportion can be written as follows:

$$\frac{8}{10} = \frac{80}{100}$$

Proportions show equivalent fractions. Both $\frac{8}{10}$ and $\frac{80}{100}$ reduce to the same fraction: $\frac{4}{5}$.

For a proportion to work, the terms in both ratios have to be written in the same order. Notice that the numerator in each ratio in the proportion above refers to the number of students expected to take the exam. The denominator refers to the total number of students.

Let's say you didn't see immediately that $\frac{8}{10}$ would be equal to $\frac{80}{100}$. How could you have figured out the equivalent ratio? Remember in Section 2 when you were working with fractions? You learned the following steps to raising a fraction to higher terms.

Step 1: Divide the denominator of the fraction into the new denominator.
Step 2: Multiply the quotient, or the answer to Step 1, by the numerator.
Step 3: Write the product, or the answer to Step 2, over the new denominator.

Let's use these steps to solve for the missing term in the proportion.

Example: $\frac{8}{10} = \frac{?}{100}$

Step 1: Divide the denominator into the new denominator. The new denominator is 100.

$100 \div 10 = 10$

Step 2: Multiply the answer to Step 1 by the numerator.

$8 \times 10 = 80$

Step 3: Write the answer to Step 2 over the new denominator.

$\frac{80}{100}$

There's another way to solve for the missing term. It's called *cross multiplying* or finding the *cross products*. Here's how cross multiplying works.

Step 1: Multiply the numerator of the first ratio by the denominator in the second ratio.

Step 2: Divide the product (the answer to Step 1) by the denominator in the first ratio. Write the answer over the denominator in the second ratio.

Let's look at the example above again and solve using cross multiplication.

Example: $\frac{8}{10} = \frac{?}{100}$

Step 1: Multiply the numerator of the first ratio by the denominator in the second ratio.

$8 \times 100 = 800$

Step 2: Divide the product (the answer to Step 1) by the denominator in the first ratio. Write the answer over the denominator in the second ratio.

$800 \div 10 = 80$

So, the second ratio is $\frac{80}{100}$.

You can also use cross multiplication to check that two ratios are equal. When a proportion is set up properly, the results of cross multiplication should be equal. Here's how it works.

EXAMPLE: Use cross multiplication to check that the two ratios in this proportion are equal:

$\frac{8}{10} = \frac{80}{100}$

continued from previous page

Step 1: Multiply the numerator of the first ratio by the denominator in the second ratio.

8 × 100 = 800

Step 2: Multiply the denominator of the first ratio by the numerator in the second ratio.

10 × 80 = 800

Step 3: Set the answer to Step 1 equal to the answer to Step 2. If they are equal, your proportion is valid.

800 = 800

Use cross multiplication to check your proportions as you work.

PRACTICE

Solve for the missing number in each proportion. Use cross multiplication to check your answers. You can check your answers at the end of the section.

17. $\frac{5}{8} = \frac{?}{32}$

18. $\frac{2}{9} = \frac{?}{810}$

19. $\frac{1}{8} = \frac{?}{640}$

20. $\frac{7}{10} = \frac{?}{110}$

21. $\frac{5}{6} = \frac{?}{36}$

22. $\frac{3}{4} = \frac{?}{120}$

23. $\frac{45}{50} = \frac{?}{250}$

24. The ratio of blue marbles to yellow marbles is 2:5. If there are 10 yellow marbles, how many blue marbles are there?

25. One inch represents 0.5 mile on a map. If the library is 3.5 inches away from your house, how many miles away is the library?

SOLVING PROPORTION WORD PROBLEMS

Proportions are common in word problems. Let's look at some examples of proportion word problems.

Example: Margaret drove 220 miles in five hours. If she maintained the same speed, how far could she drive in seven hours?

Step 1: Set up a proportion.

$$\frac{220 \text{ miles}}{5 \text{ hours}} = \frac{? \text{ miles}}{7 \text{ hours}}$$

Step 2: Solve for the missing number in the second ratio:

$220 \times 7 \div 5 = 308$

Step 3: Check your work by cross multiplying:

$220 \times 7 = 5 \times 308$

$1540 = 1540$

Keeping the same speed, Margaret could drive 308 miles in seven hours.

Example: Harry earns $6 per hour at his job. If he works nine hours this week, how much will Harry earn?

Step 1: Set up a proportion.

$$\frac{\$6}{1 \text{ hour}} = \frac{\$?}{9 \text{ hours}}$$

Step 2: Solve for the missing number in the second ratio:

$6 \times 9 \div 1 = \$54$

Step 3: Check your work by cross multiplying:

$6 \times 9 = 54 \times 1$

$54 = 54$

Working seven hours, Harry will make $54.

Example: There are 2.2 pounds in a kilogram. If a baby is born weighing 7.5 pounds, how many kilograms does the baby weigh?

Step 1: Set up a proportion.

$$\frac{2.2 \text{ pounds}}{1 \text{ kilogram}} = \frac{7.5 \text{ pounds}}{? \text{ kilograms}}$$

Step 2: Solve for the missing number in the second ratio:

$1 \times 7.5 \div 2.2 = 3.4$ (We rounded the answer to the nearest tenth of a kilogram so that our numbers were all parallel—accurate to the nearest tenth.)

Step 3: Check your work by cross multiplying:

$1 \times 7.5 = 2.2 \times 3.4$

$7.5 = 7.48$

Round 7.48 up to 7.5 to see that these are equivalent.

A baby weighing 7.5 pounds also weighs 3.4 kilograms.

PRACTICE

Solve the following problems. You can check your answers at the end of the section.

26. Sandy drove 510 miles in eight hours. If she maintained the same speed, how far could she drive in twelve hours?

27. Hermalinda earns $12 per hour. If she works 20 hours this week, how much will she earn?

28. The Jaguars have a win/lose ratio of 7:2. If they won 21 games in all, how many did they lose?

29. A gallon is equal to about 3.8 liters. How many liters are in 10 gallons?

30. There are 2.2 pounds in a kilogram. Determine your own weight in kilograms.

Use proportions to convert from one unit of measure to another. *Conversion factors* are ratios used to convert from one unit of measure to another. When you know the conversion factor, you can set up a proportion.

The units of the International System of Units, commonly referred to as SI units, are used in most countries of the world as well as by almost all scientists around the world. Sometimes SI units are called metrics. You know some of these units as centimeters, kilograms, and milliliters. Notice that many SI units have a prefix. The following table relates some SI prefixes to their meanings. You can use this table to convert between SI units.

continued from previous page

SI PREFIXES AND THEIR MEANINGS

SI PREFIX	SYMBOL	CONVERSION FACTOR	EXAMPLE OF UNIT
kilo-	k	1,000	kilometer
hecto-	h	100	hectometer
deca-	da	10	decameter
deci-	d	$\frac{1}{10}$	decimeter
centi-	c	$\frac{1}{100}$	centimeter
milli-	m	$\frac{1}{1000}$	millimeter

You can use the information in this table to convert kilometers to meters. Here's how. Based on the table, you know that a unit with the prefix kilo- (kilometer) is 1,000 larger than a unit with no prefix (meter), so 1 kilometer = 1,000 meters. Now try it with centimeters.

EXAMPLE: Convert 10 centimeters to meters.

$$10 \text{ centimeters} \times \frac{1 \text{ meter}}{100 \text{ centimeters}} = 0.1 \text{ meter}$$

So, 10 centimeters equals 0.1 meter.

You can also use conversion factors to convert between SI and conventional units. Conventional units include feet, pounds, gallons, and miles. Below are some common conversion factors used to convert between the SI and conventional systems of measurement.

1 foot = 0.3048 meter
1 inch = 2.54 centimeters
1 yard = 0.9144 meter
1 mile = 1.6 kilometers
1 gallon = 3.8 liters
1 millimeter = 0.04 inch

continued from previous page

1 centimeter = 0.3937 inch

1 kilometer = 0.62 mile

1 kilogram = 2.2 pounds

EXAMPLE: Convert 100 pounds to kilograms.

$$100 \text{ pounds} \times \frac{1 \text{ kilogram}}{2.2 \text{ pounds}} = 45.45 \text{ kilograms}$$

So, 100 pounds equals 45.45 kilograms.

Understanding Probability

LESSON SUMMARY

In this lesson, you will learn what probability is and how to calculate it. You will also practice writing probabilities as fractions, decimals, and percents.

We hear probabilities all the time. Listening to the weather report, you might hear that there is a 60% chance of rain tomorrow. At school, you might hear that 19 of 20 students will pass math this year. On TV, you might hear that $\frac{4}{5}$ of dentists recommend a certain brand of toothpaste. These are all ways of expressing probabilities.

WHAT IS PROBABILITY?

Probability is the mathematics of chance. It is a way of calculating how likely it is that something will happen. It is expressed as the following ratio:

$$\text{P (event)} = \frac{\text{Number of favorable outcomes}}{\text{Number of total outcomes}}$$

The term *favorable outcomes* refers to the events you want to occur. *Total outcomes* refers to all the possible events that could occur.

A probability of zero (0) means that the event cannot occur. A probability of 50% is said to be random or chance. A probability of 100% or 1.00 is certain to occur.

Probabilities can be written in different ways:

- As a ratio: 1 out of 2 (1:2)
- As a fraction: $\frac{1}{2}$
- As a percent: 50%
- As a decimal 0.5

Let's look at some example problems.

Example: Aili has four tickets to the School Carnival Raffle. If 150 were sold, what is the probability that one of Aili's tickets will be drawn?

Step 1: Plug the numbers into the probability equation.

$$\text{P (event)} = \frac{\text{Number of favorable outcomes}}{\text{Number of total outcomes}}$$

$$\text{P (winning ticket)} = \frac{4}{150}$$

Step 2: Solve the equation.

There are several ways to write your answer. Here are two: You can write the answer as a fraction $\frac{4}{150}$, which reduces to $\frac{2}{75}$. Or, as a percent: 2.7% (we rounded this answer up from 2.66666 . . . .).

Example: What is the probability of getting heads with a single coin toss?

Step 1: Plug the numbers into the probability equation.

$$\text{P (event)} = \frac{\text{Number of favorable outcomes}}{\text{Number of total outcomes}}$$

There are two possible outcomes: heads or tails, so you create the following equation:

$$\text{P (heads)} = \frac{1}{2}$$

Step 2: Solve the equation.

The probability of getting heads is $\frac{1}{2}$, or you have a 50% chance of getting heads.

Example: A sack holds three purple buttons, two orange buttons, and five green buttons. What is the probability of drawing a purple button out of the sack?

Step 1: Plug the numbers into the probability equation.

$$\text{P (event)} = \frac{\text{Number of favorable outcomes}}{\text{Number of total outcomes}}$$

There are 10 buttons (3 + 2 + 5 = 10) in the sack.

P (purple) = $\frac{3}{10}$

Step 2: Solve the equation.

The probability of pulling a purple button out of the sack is $\frac{3}{10}$, or there is a 30% chance of drawing a purple button out of the sack.

Example: A sack holds three purple buttons, two orange buttons, and five green buttons. What is the probability of drawing a green button out of the sack?

Step 1: Plug the numbers into the probability equation.

P (event) = $\frac{\text{Number of favorable outcomes}}{\text{Number of total outcomes}}$

There are 10 buttons (3 + 2 + 5 = 10) in the sack.

P (green) = $\frac{5}{10}$

Step 2: Solve the equation.

There is a $\frac{1}{2}$ (reduced from $\frac{5}{10}$) chance of pulling a green button out of the sack, or a 50% chance of drawing a green button out of the sack.

PRACTICE

Solve the following problems. You can check your answers at the end of the section.

1. Use the figure below to answer these questions.

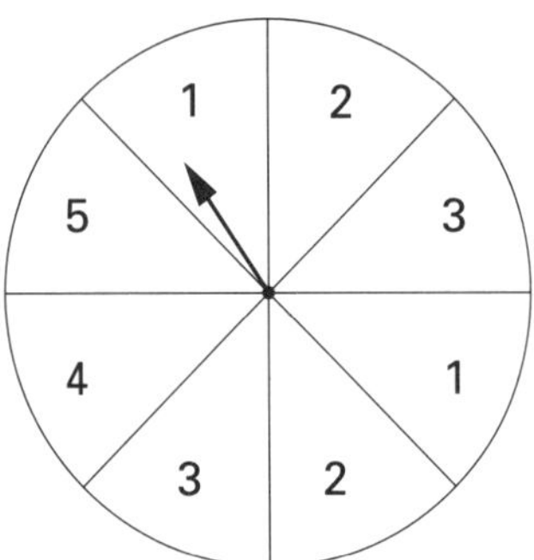

a. What is the probability of spinning a 1?
b. What is the probability of spinning an even number?
c. What is the probability of spinning a 3 or less?
d. Suppose you need to spin either a 3 or a 5 to win a game. What are the chances that you will win?
e. What are the chances of spinning a number other than 5?

2. A bag has 24 beads: 2 white, 4 purple, 6 green, and 12 pink beads.
   a. Which color bead is most likely to be selected?
   b. What is the probability of choosing a white bead?
   c. What is the probability of choosing a purple bead?
   d. What is the probability of choosing a green bead?
   e. What is the probability of choosing a pink bead?

MORE PROBABILITY PROBLEMS

So far in this lesson, you've worked with situations that deal with only one event. Now let's look at problems that involve more than one event.

Independent Events

Sometimes, the events are *independent*. That is, the first event does not affect the probability of events that come after it. Here's an example.

Example: You toss a penny and a dime into the air. What is the probability that both coins will land heads up?

You could list all the possible outcomes in a table like this:

PENNY	DIME
Heads	Heads
Heads	Tails
Tails	Heads
Tails	Tails

Then, you could use this information to fill in the probability equation:

$$\text{P (event)} = \frac{\text{Number of favorable outcomes}}{\text{Number of total outcomes}}$$

From the table, you know that there are four possible outcomes. Only one of those outcomes is heads/heads.

$$\text{P (heads/heads)} = \frac{1}{4}$$

Therefore, the probability of both coins landing heads up is $\frac{1}{4}$, or 25%.

In this problem, you had very few possible events to list. In other problems, however, you might have many possible events to account for. Another way to solve this problem is by following these steps:

Step 1: Determine the probability that each event will occur.

Step 2: Multiply the probabilities together. The product is the probability that the two events will occur.

Let's redo the last example using this method.

Example: You toss a penny and a dime into the air. What is the probability that both coins will land heads up?

Step 1: Determine the probability that each event will occur.

Each coin has two sides: heads and tails.

The probability that the penny will land heads up is $\frac{1}{2}$.

The probability that the dime will land heads up is $\frac{1}{2}$.

Step 2: Multiply the probabilities together. The product is the probability that the two events will occur together.

$\frac{1}{2} \times \frac{1}{2} = \frac{1}{4}$

Therefore, the probability of both coins landing heads up is $\frac{1}{4}$, or 25%.

Notice that the two events in the last example are independent of one another. Tossing the penny into the air does not affect the probability of outcome of tossing the dime into the air, or vice versa.

Dependent Events

Sometimes the first event does affect the probability of next event. In this case, the events are said to be *dependent*. Let's look at an example.

Example: A sack holds three purple buttons, two orange buttons, and five green buttons. What is the probability of drawing one purple button out of the sack and then—without replacing the first button—drawing a second purple button out of the sack?

Step 1: Determine the probability that each event will occur. First, notice that the first event—drawing a purple button out of the sack—affects the probability of the second event because it changes both the number of purple buttons still in the sack as well as the total number of buttons in the sack.

The probability of drawing the first purple button is $\frac{3}{10}$.

The probability of drawing a second purple button is $\frac{2}{9}$.

Step 2: Multiply the probabilities together. The product is the probability that the two events will occur together.

$$\frac{3}{10} \times \frac{2}{9} = \frac{6}{90} = \frac{1}{15}$$

Therefore, the answer is $\frac{1}{15}$, or 6.7%.

PRACTICE

Solve the following problems. You can check your answers at the end of the section.

3. In a dice game, you roll two six-sided dice. What is the probability that the dice will total 2?

4. Look at the spinner below. What are the chances of spinning 5 twice in a row?

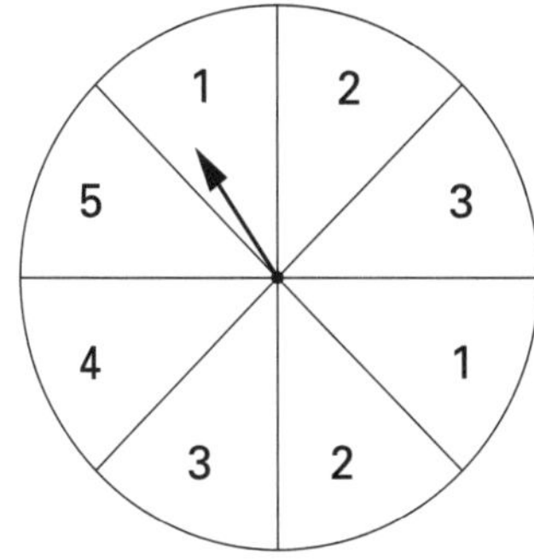

5. Assume that when you grow up, you plan to have two children. (Assume that you will not have identical twins.) What is the probability that both of your children will be girls?

6. A woman is pregnant with her second child. Her first child is a boy. What is the probability that her second child will also be a boy?

7. You have ten cards in your hand: 5 hearts and 5 spades. If your friend takes and keeps two cards from your hand, what is the probability that both cards are spades?

8. Each person in your class of 30 students writes their name on a slip of paper and puts it in a box. Your teacher will draw out two names—these students will represent your class on the Student Council this month.

a. What is the probability that your name will be drawn second?

b. What is the probability that you are selected first and your best friend is selected second?

Real World Problems

These problems apply the skills you've learned in Section 5 to everyday situations. As you work through these problems, you'll see that the skills you've learned in this section aren't only important for math tests, they are important skills for ordinary questions that come up every day.

After you've arrived at an answer for a problem, ask yourself if the answer makes sense. For example, the answer to a problem asking for the mean, median, or mode should always be in between the lowest and highest numbers in the series given. If your answer is outside the range, you should take another look at your answer.

1. During the last week of track training, Shoshanna achieves the following times in seconds: 66, 57, 54, 54, 64, 59, and 59. Her three best times this week are averaged for her final score on the course. What is her final score?
2. If a vehicle is driven 22 miles on Monday, 25 miles on Tuesday, and 19 miles on Wednesday, what is the average number of miles driven each day?

3. The chart below lists the number of students present at the monthly meetings for the Environment Protection Club. What was the average monthly attendance over the course of all the months listed?

MONTH	NUMBER OF STUDENTS PRESENT
September	54
October	61
November	70
December	75

4. Rita made a list of all of her relatives and then counted up the number of people in each category. She counted 7 aunts, 8 uncles, 4 grandparents, 22 cousins, 3 siblings, and 4 nieces. What is the mode of these numbers?

5. Stephen recorded the number of butterflies he saw in his backyard for four months and put the information in the table below. What was the mean number of butterflies in Stephen's backyard during these four months?

MONTH	NUMBER OF BUTTERFLIES SPOTTED
May	28
June	44
July	64
August	56

6. The Burnsville Neighborhood Association has been growing for the past three years. Its membership was 486 in 1999, 591 in 2000, and 573 in 2001. What is the mean membership for these three years?

7. Lefty keeps track of the length of each fish that he catches. These are the lengths in inches of the fish that he caught one day: 12, 13, 8, 10, 8, 9, 17. What is the median fish length that Lefty caught that day? What is the mode of the fish lengths he caught that day?

8. Mary Beth works in the men's department at a local department store. Yesterday, she sold men's pants in following waist sizes: 40, 32, 34, 31, 36, 34, 33, 30, 28, 34. What is the mean, median, and mode of the waist sizes sold yesterday? Which of these statistics will probably be most valuable to Mary Beth's manager?

9. At the Knott Block Party, two drawings were held for a new rose bush and a pair of gardening gloves. A jar held the names of 26 different people, and without looking, Mrs. Fikstad, the oldest woman on the block, drew a name.

   a. After picking Benjamin's name for the rose bush, Mrs. Fikstad replaced it in the jar and mixed up the names. What is the probability that Mrs. Fikstad then selected Benjamin's name to win the pair of gardening gloves?

   b. After picking Benjamin's name for the rose bush, Mrs. Fikstad set it aside and mixed up the remaining names. What is the probability that Mrs. Fikstad then selected Francine's name to win the pair of gardening gloves?

10. The number of red blood cells in one cubic millimeter of human blood is about 5,000,000. The number of white blood cells in one cubic millimeter of human blood is about 8,000. What, then, is the ratio of white blood cells to red blood cells in human blood?

11. A recipe serves six people and calls for $2\frac{1}{4}$ cups of broth. If you want to serve four people, how much broth do you need?

12. On a trip to France, you notice that gasoline costs 7 French francs per liter. You know that a gallon is equal to 3.8 liters and one dollar is equal to 7.5 French francs. How much does the gas cost in dollars per gallon?

13. A bag of jellybeans contains 8 black beans, 10 green beans, 3 yellow beans, and 9 orange beans.
   a. What is the probability of selecting either a yellow or an orange bean?
   b. What is the probability of selecting a yellow or an orange bead, taking that bead out of the jar and not replacing it, and then selecting a green bead?

14. In Derrick's sock drawer, there were 5 blue socks, 12 white socks, and 7 black socks. He reached into the drawer without looking and pulled out the first sock he touched. What were the odds he pulled out a blue sock?

SECTION

5

Answers & Explanations

LESSON 11

1. The mean is the same as the average of the numbers, so you add all the numbers together and divide by 7, the total amount of numbers:

$$\frac{3 + 7 + 8 + 10 + 3 + 5 + 6}{7} = 6$$

So, the mean is 6.

2. The mean is the same as the average of the numbers, so you add all the numbers together and divide by 5, the total amount of numbers:

$$\frac{23 + 45 + 67 + 48 + 36}{5} = 43.8$$

So, the mean is 43.8.

3. The mean is the same as the average of the numbers, so you add all the numbers together and divide by 5, the total amount of numbers:

$$\frac{88 + 92 + 100 + 95 + 85}{5} = 92$$

So, the mean is 92.

4. The mean is the same as the average of the numbers, so you add all the numbers together and divide by 8, the total amount of numbers:

$$\frac{3 + 4 + 5 + 6 + 7 + 8 + 9 + 6}{8} = 6$$

So, the mean is 6.

5. The mean is the same as the average of the numbers, so you add all the numbers together and divide by 6, the total amount of numbers:

$$\frac{100 + 105 + 110 + 101 + 103 + 107}{6} = 104.33$$

So, the mean is 104.33.

6. The mean is the same as the average of the numbers, so you add all of Misha's grades together and divide by 10, the total amount of grades she received:

$$\frac{82 + 85 + 72 + 90 + 88 + 86 + 91 + 93 + 81 + 85}{10} = 85.3$$

Thus, Misha's average in English is 85.3, or rounded, 85.

When dividing by a multiple of 10, you just have to move the decimal point.

7. The mean is the same as the average of the numbers, so you add all Jack's grades together and divide by 3:

$$\frac{82 + 81 + 84}{3} = 82.3$$

Because he needs an 83 to get a B, Jack does not have a B in Math so far.

8. The mean is the same as the average of the numbers, so you add all Mara's weekend earnings together and divide by 4, the number of weekends Mara babysat:

$$\frac{\$25 + \$27 + \$13 + \$18}{4} = \$20.75$$

Mara averaged $20.75 per weekend of babysitting.

9. You want to find the average of the numbers, so you add all the amounts John spent on each brother and divide by 3, since John has three brothers:

$$\frac{\$21 + \$17 + \$19}{3} = \$19$$

On average, John spent $19 per brother on birthday presents last year.

10. You want to find the average of the numbers, so you add all the amounts Philip saved in a week and divide by the number of weeks, 4:

$$\frac{\$5 + \$7 + \$2 + \$9}{4} = \$5.75$$

On average, Philip saved $5.75 per week from his allowance.

11. First, put the numbers in sequential order: 3, 3, 5, 6, 7, 8, 10. The middle number is the median, so the answer is 6.

12. First, put the numbers in sequential order: 23, 36, 45, 48, 67. The middle number is the median, so the answer is 45.

13. First, put the numbers in sequential order: 85, 88, 92, 95, 100. The middle number is the median, so the answer is 92.

14. The numbers are already in sequential order: 3, 4, 5, 6, 7, 8, 9, 10, 11. The middle number is the median, so the answer is 7.

15. First, put the numbers in sequential order: 100, 101, 103, 104, 105, 107, 110. The middle number is the median, so the answer is 104.

16. First, put the numbers in sequential order: 11, 12, 14, 16, 17. The middle number is the median, so the answer is 14.

17. First, put the numbers in sequential order: 3, 3, 4, 7, 8, 9. There are two middle numbers: 4 and 7. So, you have to take the average of these two numbers to determine the median: $(4 + 7) \div 2 = 5.5$. So, the median is 5.5.

18. First, put the numbers in sequential order: 5, 6, 7, 8, 9, 10. There are two middle numbers: 7 and 8. So, you have to take the average of these two numbers to determine the median: $(7 + 8) \div 2 = 7.5$. So, the median is 7.5.

19. First, put the points scored in sequential order: 1, 2, 3, 5, 6. The middle number is the median: 3. Jill's median points scored over the last five games is 3.

20. First, put Jack's grades in sequential order: 81, 82, 84. The middle number is the median. So, Jack's median grade in Math is 82.

21. The mode is 3 because it appears in the set of numbers most frequently.

22. The mode is 45 because it appears in the set of numbers most frequently.

23. The mode is 5 because it appears in the set of numbers most frequently.

24. No numbers occur more than once. There is no mode.

25. The mode is 103 because it appears in the set of numbers most frequently.

26. Two numbers occur more than once, so there are two modes: 12 and 17.

27. Two numbers occur more than once, so there are two modes: 3 and 7.

28. No numbers occur more than once, so there is no mode.

29. Two numbers occur more than once, so there are two modes: 25 and 39.

30. Three numbers occur more than once, so there are three modes: 3, 7, and 8.

LESSON 12

1. You just have to set the comparison up in a fraction, so you get $\frac{12}{15}$, which can be reduced to $\frac{4}{5}$.

2. You just have to set the comparison up in a fraction, so you get $\frac{15}{20}$, which can be reduced to $\frac{3}{4}$.

3. You just have to set the comparison up in a fraction, so you get $\frac{5}{10}$, which can be reduced to $\frac{1}{2}$.

4. You just have to set the comparison up in a fraction, so you get $\frac{6}{4}$, which can be reduced to $\frac{3}{2}$.

5. You just have to set the comparison up in a fraction, so you get $\frac{2}{1}$. Remember not to change the improper fraction to a mixed number. A ratio that looks like a fraction is comparing the two numbers, and you will lose the comparison if you treat the ratio as a fraction and convert it to a mixed number.

6. You just have to set the comparison up in a fraction, so you get 5:40, which can be reduced to 1:8.

7. You just have to set the comparison up in a fraction, so you get 1:1.

8. You just have to set the comparison up in a fraction, so you get 3:9, which can be reduced to 1:3.

9. You just have to set the comparison up in a fraction, so you get 3:1. Remember not to change the improper fraction to a mixed number. A ratio that looks like a fraction is comparing the two numbers, and you will lose the comparison if you treat the ratio as a fraction and convert it to a mixed number.

10. You just have to set the comparison up in a fraction, so you get 2:10, which can be reduced to 1:5.

11. Marc got 21 questions correct and 4 incorrect. Therefore, the ratio is 21:4.

12. First, calculate the total number of questions on the test: 21 + 4 = 25. Then, write the number of questions Marc got right over the total number of questions on the test: $\frac{21}{25}$.

13. First, write the ratio: $\frac{325 \text{ miles}}{5 \text{ hours}}$. Then reduce by dividing by 5 and you get $\frac{65 \text{ miles}}{\text{hour}}$. So, the driver drove, on average, 65 miles per hour.

14. This ratio will involve more than two numbers, so you cannot write it as a fraction. Write the numbers in the order that they are requested: 5 dogs:10 cats:15 chickens:20 hamsters. Reduce the terms by dividing each number by 5. 1 dog:2 cats:3 chickens:4 hamsters.

15. Write the numbers in the order requested: 5 dogs to 10 cats. Reduce to lowest terms by dividing each number by 5. 1 dog:2 cats.

16. First, calculate the total number of animals that participated in the show: 15 + 5 + 20 + 10 = 50 animals. Then, write the number of cats over the total number of animals and reduce to lowest terms: $\frac{10 \text{ cats}}{50 \text{ total animals}}$ reduces to $\frac{1}{5}$.

17. $\frac{5}{8} = \frac{20}{32}$

Cross multiply to check your work:

$5 \times 32 = 820$

$160 = 160$

18. $\frac{2}{9} = \frac{180}{810}$

Cross multiply to check your work:

$2 \times 810 = 9 \times 180$

$1{,}620 = 1{,}620$

19. $\frac{1}{8} = \frac{80}{640}$

Cross multiply to check your work:

$1 \times 640 = 880$

$640 = 640$

20. $\frac{7}{10} = \frac{77}{110}$

Cross multiply to check your work:

$7 \times 110 = 10 \times 77$

$770 = 770$

21. $\frac{5}{6} = \frac{30}{36}$

Cross multiply to check your work:

$5 \times 36 = 6 \times 30$

$180 = 180$

22. $\frac{3}{4} = \frac{90}{120}$

Cross multiply to check your work:

$4 \times 90 = 3 \times 120$

$360 = 360$

23. $\frac{45}{50} = \frac{225}{250}$

Cross multiply to check your work:

$45 \times 250 = 50 \times 225$

$11{,}250 = 11{,}250$

24. First, set up the proportion:

$$\frac{\text{2 blue marbles}}{\text{5 yellow marbles}} = \frac{\text{? blue marbles}}{\text{10 yellow marbles}}$$

Then solve for the missing number using one of the methods in the lesson. One way is to multiply 2 and 10 and divide by 5, which yields 4. So there are 4 blue marbles. Check your work using cross multiplication:

$2 \times 10 = 54$

$20 = 20$

25. First, set up the proportion:

$$\frac{\text{1 inch}}{\text{0.5 mile}} = \frac{\text{3.5 inches}}{\text{? miles}}$$

Then solve for the missing number using one of the methods in the lesson. One way is to multiply 0.5 and 3.5 and divide by 1, which yields 1.75. So the library is 1.75 miles from home. Check your work using cross multiplication:

$1 \times 1.75 = 0.5 \times 3.5$

$1.75 = 1.75$

26. Begin by setting up a proportion.

$$\frac{\text{510 miles}}{\text{8 hours}} = \frac{\text{? miles}}{\text{12 hours}}$$

Then, solve for the missing number in the second ratio: $510 \times 12 \div 8 = 765$. Check your work by cross multiplying:

$510 \times 12 = 8 \times 765$

$6{,}120 = 6{,}120$

Keeping the same speed, Sandy could drive 765 miles in 12 hours.

27. Begin by setting up a proportion.

$$\frac{\$12}{\text{1 hour}} = \frac{\$?}{\text{20 hours}}$$

Then, solve for the missing number in the second ratio: $12 \times 20 \div 1 = 240$. Check your work by cross multiplying:

$12 \times 20 = 1 \times 240$

$240 = 240$

Working 20 hours, Hermalinda will make $240.

28. Begin by setting up a proportion.

$$\frac{7 \text{ wins}}{2 \text{ losses}} = \frac{21 \text{ wins}}{? \text{ losses}}$$

Then, solve for the missing number in the second ratio: $2 \times 21 \div 7 = 6$. Check your work by cross multiplying:

$$2 \times 21 = 6 \times 7$$
$$42 = 42$$

They lost 6 games in all.

29. Begin by setting up a proportion.

$$\frac{1 \text{ gallon}}{3.8 \text{ liters}} = \frac{10 \text{ gallons}}{? \text{ liters}}$$

Then, solve for the missing number in the second ratio: $3.8 \times 10 \div 1 = 38$. Check your work by cross multiplying:

$$3.8 \times 10 = 1 \times 38$$
$$38 = 38$$

Ten gallons equals 38 liters.

30. Answers will depend on your weight. Follow the same steps shown for problems 26–29 to calculate your weight in kilograms.

LESSON 13

1. **a.** There are two 1s and eight possible outcomes. Therefore, the probability of spinning a 1 is $\frac{2}{8}$, $\frac{1}{4}$, or 25%.

b. First, count the even numbers on the spinner: 2, 2, and 4 are all even. There are three even numbers and eight possible outcomes. Therefore, the probability of spinning an even number is $\frac{3}{8}$ or 37.5%.

c. First, count the numbers that are equal to or less than 3: there are six. There are eight possible outcomes, so the probability of spinning a 3 or less is $\frac{6}{8}$, $\frac{3}{4}$, or 75%.

d. There are two 3s and one 5. There are three possible favorable outcomes and eight total possible outcomes. The probability of spinning a 3 or a 5 is $\frac{3}{8}$ or 37.5%.

e. First, count all the numbers on the spinner that are not 5. That's all the numbers except the one section labeled 5. Therefore, the probability of spinning a number other than 5 is $\frac{7}{8}$ or 87.5%.

2. **a.** There are more pink beads than any other color, so the chances of drawing a pink bead should be greater than drawing any other specific color bead.

b. There are 2 white beads and 24 possible outcomes. Therefore, the probability of drawing a white bead is $\frac{2}{24}$, $\frac{1}{12}$, or 8.3%.

c. There are 4 purple beads and 24 possible outcomes. Therefore, the probability of drawing a purple bead is $\frac{4}{24}$, $\frac{1}{6}$, or 16.7%.

d. There are 6 green beads and 24 possible outcomes. Therefore, the probability of drawing a green bead is $\frac{6}{24}$, $\frac{1}{4}$, or 25%.

e. There are 12 pink beads and 24 possible outcomes. Therefore, the probability of drawing a pink bead is $\frac{12}{24}$, $\frac{1}{2}$, or 50%.

3. These are independent events. In order for the dice to total 2, each one will need to land with a 1 face up. Each die has six sides and only one side has a 1 on it. So the probability of the first die landing with a 1 is $\frac{1}{6}$. The probability of the second die landing with a 1 on it is also $\frac{1}{6}$. The probability of rolling a 2 is $\frac{1}{6} \times \frac{1}{6} = \frac{1}{36}$.

4. These are independent events. The probability of spinning a 5 the first time is $\frac{1}{8}$. The probability of spinning 5 the second time is also $\frac{1}{8}$. So the probability of rolling a 5 twice in a row is $\frac{1}{8} \times \frac{1}{8} = \frac{1}{64}$.

5. These are independent events. The probability of having a girl the first time is $\frac{1}{2}$ (there are two possible outcomes: girl or boy). The probability of having a girl the second time is also $\frac{1}{2}$. So the probability of having two girls is $\frac{1}{2} \times \frac{1}{2} = \frac{1}{4}$, or 25%.

6. These are independent events, but the first event has already occurred. The probability of having a boy the first time is 1, or 100%. (The child is already born, so the event is a certainty.) The probability of having a boy the second time is $\frac{1}{2}$, or 50%. So the probability of having a boy the second time is $1 \times \frac{1}{2} = \frac{1}{2}$, or 50%.

7. These are dependent events. The probability of drawing a spade the first time is $\frac{5}{10}$, or $\frac{1}{2}$. The probability of drawing a spade the second time is $\frac{4}{9}$. So the probability of drawing two spades is $\frac{1}{2} \times \frac{4}{9} = \frac{4}{18} = \frac{2}{9}$.

8. a. This is only one event. There are 29 names left in the box for the second drawing, so the probability that your teacher will select your name is $\frac{1}{29}$.

b. These are dependent events. The probability that you or your best friend will be selected first is $\frac{1}{30}$. The probability that the other of you will be selected the second time is $\frac{1}{29}$. So the probability that you will both be selected is $\frac{1}{30} \times \frac{1}{29} = \frac{1}{870}$—not likely.

▶ REAL WORLD PROBLEMS

1. Shoshanna's final score is the average of her three best times. Her three best times are 54, 54, and 57 seconds. To average these times, add them all together and divide by three: (54 + 54 + 57)/3. Her final score is 55 seconds.

2. Begin by adding the miles driven each day together: Then divide by three. The average number of miles driven each day is 22.

3. Begin by adding the number of students attending the meeting each month. Then divide by the number of months shown (4). The average monthly attendance over the course of the months listed is 65 students.

4. Rita has four grandparents and four cousins. Four is the only number that recurs in the list. The mode is 4.

5. Add up the total number of butterflies seen each month and divide by the number of months (4). The mean number of butterflies in Stephen's backyard during these four months is 48.

6. Add up the total membership over the three years and divide by 3. The mean membership for the three years is 550.

7. To determine the median, first put the numbers in sequential order: 8, 8, 9, 10, 12, 13, 17. The middle number is the median, so the median is 10. The number 8 occurs twice in the series, so the mode is 8.

8. To calculate the mean, add up all the numbers in the list and divide by 10. The mean is 33.2 inches. To calculate the median, put the numbers in sequential order: 28, 30, 31, 32, 33, 34, 34, 34, 40, 36. There are two middle numbers 33 and 34. The median is the average of the two: 33.5. The number 34 occurs three times. No other number recurs in the list. So the mode is 34. The mode will be the most valuable of these three statistics because it will tell the manager what size of pants is more likely to sell and, thus, which size to stock in the store.

9. a. These are independent events. Benjamin's name occurs once in the jar. There are 26 possible names in the jar. So the probability that Benjamin's name is chosen is $\frac{1}{26}$.

b. Now that Benjamin's name has been removed from the jar, there are only 25 possible names to draw. Francine's name occurs only once, so the probability of drawing Francine's name for the gardening gloves is $\frac{1}{25}$.

10. Set up a ratio as follows:

$$\frac{8{,}000 \text{ white blood cells}}{5{,}000{,}000 \text{ red blood cells}}$$

Then, reduce the ratio (by dividing by 8,000) to 1:625.

11. Begin by setting up a proportion.

$$\frac{6 \text{ people}}{2\frac{1}{4} \text{ cups of broth}} = \frac{4 \text{ people}}{? \text{ cups of broth}}$$

Then, solve for the missing number in the second ratio: $4 \times 2\frac{1}{4} \div 6 = 1\frac{1}{2}$. Check your work by cross multiplying:

$$4 \times 2\frac{1}{4} = 6 \times 1\frac{1}{2}$$
$$9 = 9$$

You need $1\frac{1}{2}$ cups of broth to serve four people.

12. This problem requires setting up and solving two proportions. The order is not important. You might first determine the cost of a gallon of gas in French francs (FF).

$$\frac{7 \text{ FF}}{1 \text{ liter}} = \frac{? \text{ FF}}{3.8 \text{ liters}}$$

Notice that you use 3.8 liters instead of one gallon because you must use the same units in both proportions. Then, solve for the missing number in the second ratio: $7 \times 3.8 \div 1 = 26.6$. Now you know that one gallon of gas costs 26.6 FF in France. Next you can calculate the cost in dollars.

$$\frac{\$1}{7.5\text{ FF}} = \frac{\$\,?}{26.55\text{ FF}}$$

Then, solve for the missing number in the second ratio: $1 \times 26.6 \div 7.5 = 3.55$. So, one gallon of gas in France costs \$3.55.

13. **a.** There are three yellow and nine orange jelly beans, so there are 12 possible favorable outcomes ($3 + 9 = 12$). There are total of 30 jellybeans in the jar. So there is a $\frac{12}{30}$, or $\frac{2}{5}$, chance of selecting either a yellow or an orange jellybean from the jar.

b. These are dependent events. The probability of drawing either a yellow or an orange jellybean the first time is $\frac{2}{5}$. The probability of drawing a green jellybean the second time is $\frac{10}{29}$. So the probability of drawing a yellow or an orange jellybean, taking that jellybean out of the jar, and then drawing a green jellybean is $\frac{2}{5} \times \frac{10}{29} = \frac{20}{145} = \frac{4}{29}$.

You can use canceling to reduce the fraction to lowest terms.

$$\frac{2}{\cancel{5}_{1}} \times \frac{\cancel{10}^{2}}{29} = \frac{4}{29}$$

14. There are five blue socks and a total of 24 socks in the drawer. So the odds are $\frac{5}{24}$ that Derrick pulled out a blue sock.

SECTION

6

Graphs, Tables, and Charts

IN CLASS, ON tests, and in your day-to-day life you see graphs, tables, and charts everywhere. They organize information visually, so that you can *see* what it looks like instead of having a list of numbers. For example, when you watch a baseball game on TV, television sports producers flash tables and charts on the screen so viewers can absorb information about a team's standings in the league or a player's averages at a glance and quickly return to the game. The goal of graphs, tables, and charts is to explain a problem, situation, or opportunity so viewers can draw a conclusion, find a solution, or develop a plan. An effective chart finds relationships that were hidden behind the raw numbers or pages of text on which the chart was based.

Interpreting Pie Charts, Line Graphs, and Bar Graphs

LESSON SUMMARY

In this lesson, you will learn three ways data is shown visually in graphs and charts. You'll learn how to read and interpret pie charts, line graphs, and bar graphs as well as how to decide when one type of chart or graph is better suited to a set of data than another type of graph.

When you pick up the newspaper or watch a news report on TV, you'll often see information presented in a graph. More and more, we give and receive information visually. That's one reason you're likely to find graphs on math tests, and a good reason to understand how to read them. Let's look at three common kinds of graphs—pie charts, line graphs, and bar graphs.

PIE CHARTS

Pie charts show how the parts of a whole relate to one another. A pie chart is a circle divided into slices or wedges—like a pizza. Each slice represents a category. Pie charts are sometimes called circle graphs. Let's look at some examples of pie charts and see what kinds of information they can provide.

Example: Susie is a student representative on the computer lab committee. The committee wants to build a new computer lab at Martin Luther King, Jr., Middle School. The committee needs to raise $50,000. The pie chart below shows how they plan to spend the money. Use the pie chart to answer these questions.

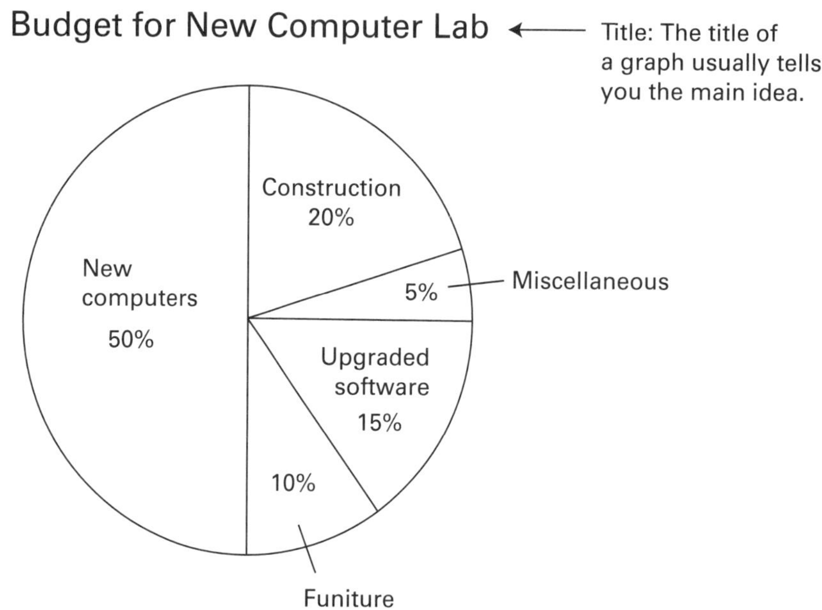

a. How much of the money does the committee plan to spend on new computers?
b. What percent of the budget is to be spent on new computers and upgraded software?
c. What percent of the budget will *not* be spent on new computers or upgraded software?

Explanations:

a. The committee wants to raise $50,000. According to the pie chart, 50%, or half, of this money will be spent on new computers. So, $25,000 is budgeted for new computers.

b. According to the pie chart, 50% of the budget will go toward the purchase of new computers and 15% of the budget will go toward the purchase of upgraded software. Therefore, the committee plans to use 65% (50% + 15%) of the budget to buy new computers and software.

c. There are two ways of approaching this question. Since you already know that 65% of the budget is set aside for new computers and upgraded software, you could subtract 65% from 100% to get 35%. Or, you could add up the other items in the pie chart:

Construction	20%
Furniture	10%
+Miscellaneous	5%
	35%

Using either method, the answer is 35%.

PRACTICE

Use each graph to answer the questions that follow it. You can check your answers at the end of the section.

1. Nicholas found the following pie chart in his Physical Science textbook.

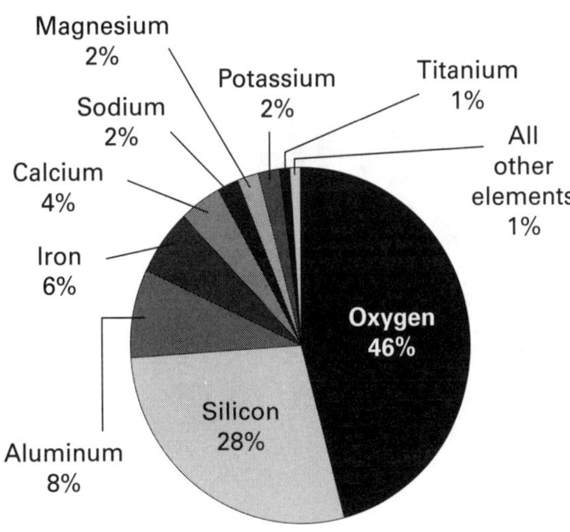

a. Which single element makes up the largest percent of Earth's crust?
b. Which two elements make up most of Earth's crust? What percent of the crust do these two elements represent together?
c. Together, what percent of Earth's crust do iron and aluminum make up?

2. Jackie found the following pie charts in the newspaper. They represent data collected from a recent telephone survey.

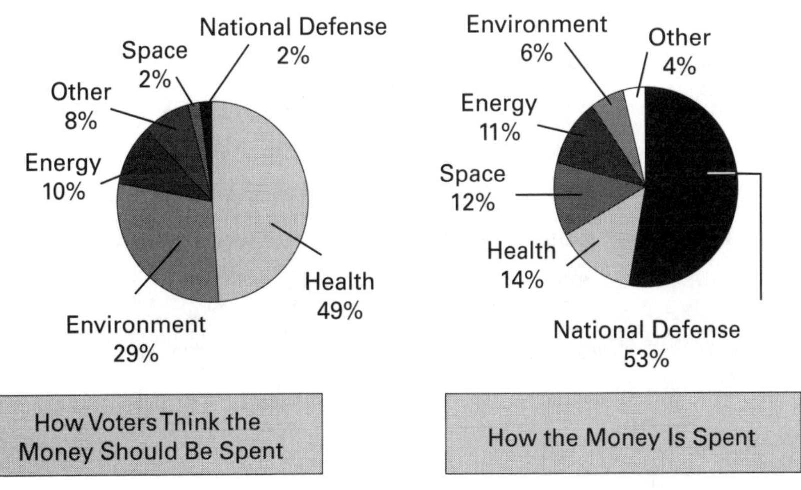

a. Based on the survey, which category of spending best matches the voter's wishes?
b. On which category of spending did the voters say they want most of the money spent?
c. Which category of spending receives the most federal dollars?
d. To which two categories of spending did voters say they wanted the most money to go? Which two categories of spending actually received the most money?

3. The Johnsons made the following graph of their monthly family budget.

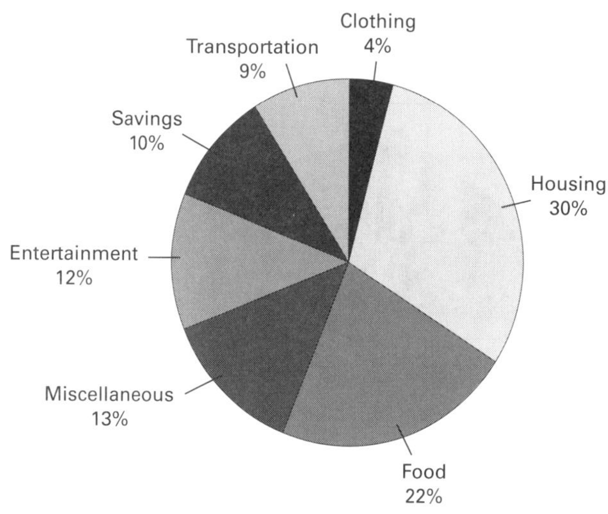

a. Where does most of the budget go each month?
b. What is the smallest category of spending in the budget?
c. In percent of overall expenses, how much more money is spent on food than on transportation and clothing combined?
d. If the Johnsons make $2,000 each month, how much do they save each month?

LINE GRAPHS

Line graphs show how two categories of data or information (sometimes called *variables*) relate to one another. The data is displayed on a grid and is presented on a scale using a horizontal and a vertical axis for the different categories of information the graph is comparing. Usually, each data point is connected together to form a line so that you can see trends in the data, or how the data changes over time. Therefore, often you will see line graphs with *time* on the horizontal axis. Let's look at an example of a line graph and see what kinds of information it can provide.

Example: Julia was very excited when her baby sister was born. She made the following line graph to chart her sister's weight gain over the first year. Use the line graph to answer the following questions.

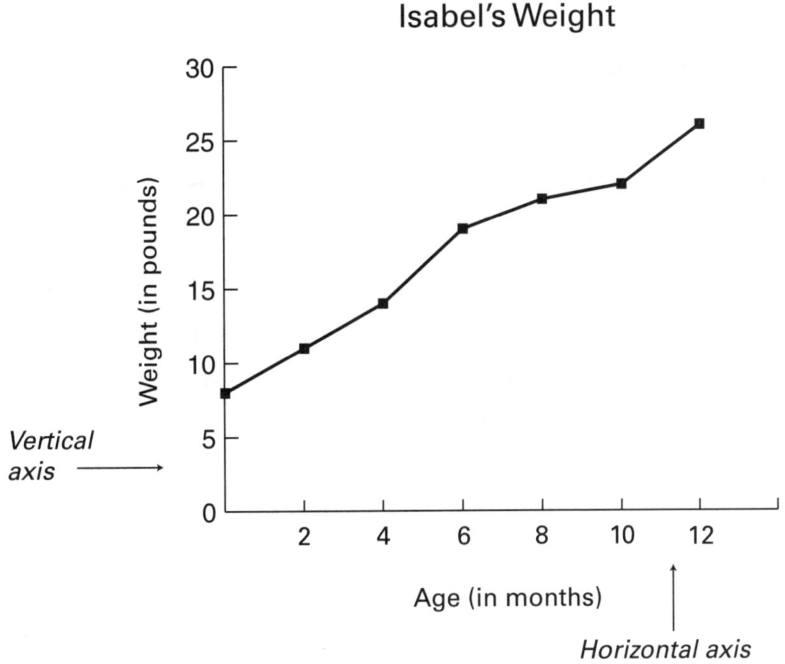

a. What variable is shown on the vertical axis? What variable is shown on the horizontal axis?
b. During which two-month period did Isabel have the greatest weight gain?
c. During which two-month period did Isabel have the smallest weight gain?
d. The average weight for a one-year-old girl is 20 pounds. How much more than average did Isabel weigh at one year?
e. How much weight did Isabel gain during her first year?

Explanations:

a. Look at the labels. Isabel's weight in pounds is shown on the vertical axis. Her age in months is shown on the horizontal axis.
b. Look at how steep the line is. The steepest part of the line represents the fastest weight gain. The line is the steepest between 10 and 12 months. Isabel gained the most weight during this time.
c. Look at how steep the line is. The flattest part of the line represents the slowest weight gain. The line is flattest between 8 and 10 months. Isabel gained the least weight during this period.
d. According to the graph, Isabel weighed about 26 pounds at 12 months. Subtract the average weight from this amount:

$26 - 20 = 6$

Isabel weighed about 6 pounds more than the average girl at twelve months.

e. Isabel weighed about 8 pounds at birth. She weighed about 26 pounds at 12 months. Subtract her weight at birth from her weight at 12 months:

26 – 8 = 18

Isabel gained about 18 pounds during her first year.

Line graphs are often used to show the results of a scientific experiment. The variable that the scientist is measuring and tracking is often called the *dependent variable*. It is usually measured on the vertical axis of a graph. The horizontal axis is usually measuring time, so you can see how the data changes over time.

PRACTICE

Use each graph to answer the questions that follow it. You can check your answers at the end of the section.

4. As part of his science project, Edwin measures the height of a plant every other day and marks it on a line graph. Edwin's graph is shown below.

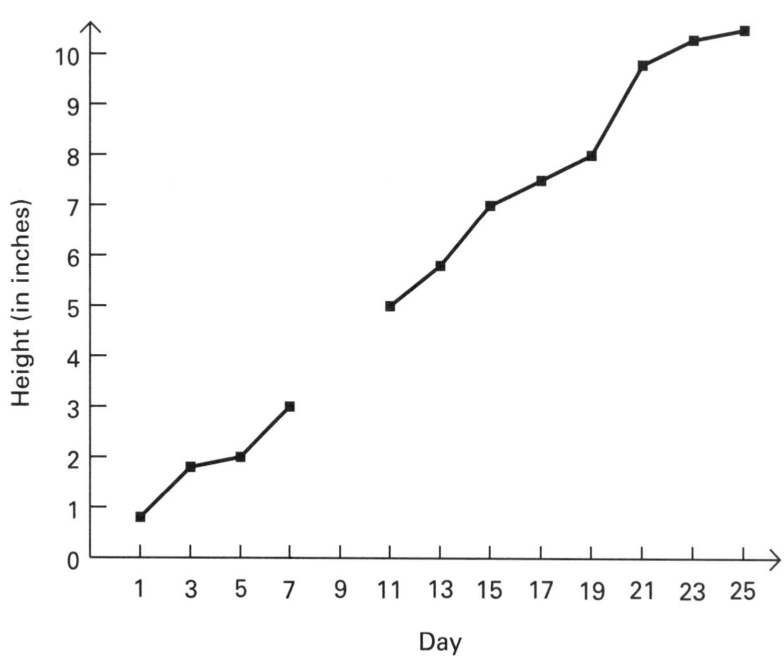

a. What variable is shown on the vertical axis? What variable is shown on the horizontal axis?

b. Edwin forgot to measure the height of the plant on day 9. What is the most likely height of the plant on that day?

c. During which two-day period did the plant grow the most?

d. During which two-day period did the plant grow the least?

e. How much did the plant grow from day 1 to day 25?

5. Jeannie saw the following graph on the news last night.

a. What variable is shown on the vertical axis? What variable is shown on the horizontal axis?

b. Based on the graph, at what density are people most likely to use public transportation?

c. Based on the graph, at what density are people most likely to use their own car to get to work?

d. Which form of transportation becomes less popular as population density increases?

▶ BAR GRAPHS

Like pie charts, *bar graphs* show how different categories of data relate to one another. A bar represents each category. The length of the bar represents the relative frequency of the category—compared to the other categories on the graph. Let's look at an example of a bar graph and see what kinds of information it can provide.

Both pie charts and bar graphs are used to compare different categories of data. So when you have data to graph, how do you decide which kind of graph to use? Think about what your purpose is. If your purpose is to compare the absolute values of each category, then a bar chart is probably better because the amounts of each category are shown in comparison to each other. If your purpose is to show how each part relates to the whole, a pie chart is probably better.

Example: Vaughan's teacher asked him to compare the rainfall in Cherokee County this year with the average rainfall in Cherokee County over the last five years. He constructed the following graph using data from his local library. Use the bar graph to answer the following questions.

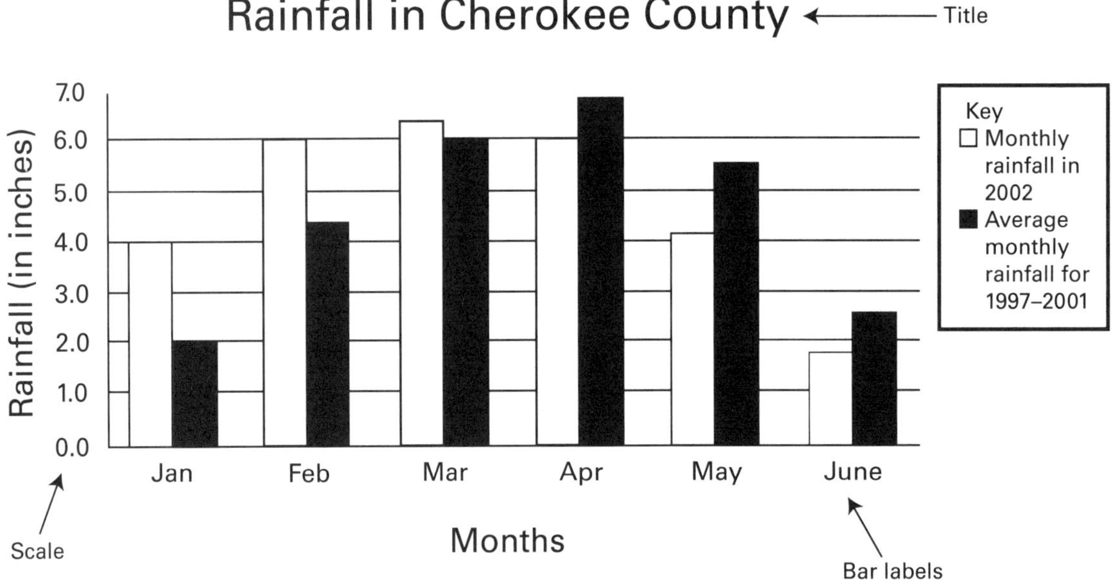

a. What does each bar represent? What is the difference between the shaded bars and the white bars?

b. During which months is the rainfall in 2002 greater than the average rainfall?

c. During which months is the rainfall in 2002 less than the average rainfall?

d. How many more inches of rain fell in April 2002 than in January 2002?

e. How many more inches of rain fell in January 2002 than on average during the last five years in January?

Explanations

a. Look at the labels and the key. Each bar represents the number of inches of rainfall during a particular month. From the key, you know that the shaded bars represent the average monthly rainfall for 1997–2001. The white bars represent the rainfall in 2002.

b. Compare the white bars with the shaded bars. Rainfall in 2002 is greater than average during the months that the white bar is taller than the shaded bar for that month. Rainfall in 2002 was greater than the average rainfall during January, February, and March.

c. Compare the white bars with the shaded bars. Rainfall in 2002 is less than average during the months that the shaded bar is taller than the white bar for that month. Rainfall in 2002 was less than the average rainfall during April, May, and June.

d. Compare the height of the white bars for January and April. In April, 6 inches of rain fell. In January, 4 inches of rain fell. Then subtract: 6 – 4 = 2. So, In April, 2 more inches of rain fell than in January.

e. Compare the height of the shaded bar and the white bar for January. The shaded bar represents 2 inches. The white bar represents 4 inches. Subtract: 4 – 2 = 2. So, two more inches of rain fell in January 2002 than on average during the last five years in January.

The bar chart in the last example compares the amount of rainfall over time. This could have been shown on a line graph. In fact, line graphs often show the change in one variable over time. So how do you decide when to use a bar graph and when to use a line graph? Again, it depends on your purpose. If you want to emphasize the trend—how the variable changed over time—then a line graph is better. If you want to compare the relative sizes of each category, a bar chart is better. Also, if you have many categories, a line graph will be easier for your reader to read than a bar graph.

PRACTICE

Use each graph to answer the questions that follow it. You can check your answers at the end of the section.

6. Emmanuel used data from a local pet store to compare the amounts of dog food different breeds eat in one month. He recorded his data in the following graph.

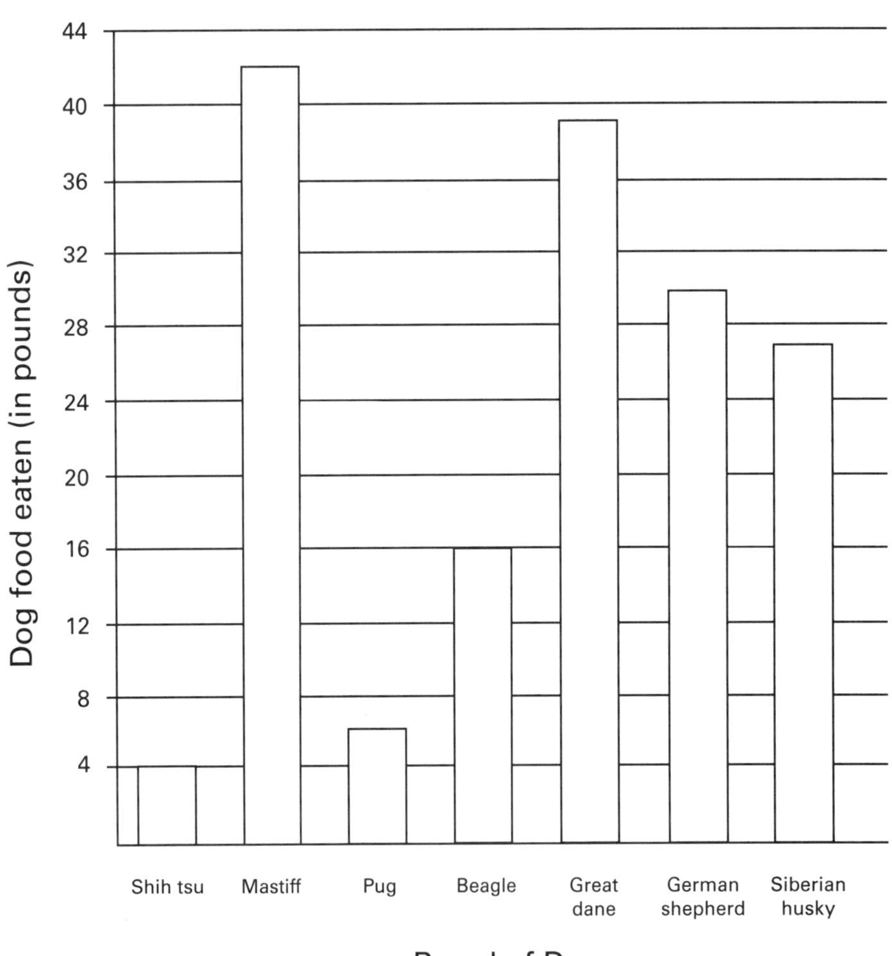

a. What does each bar represent?

b. Which breed of dog ate the most food during the month that Emmanuel graphed? How many pounds of food did this breed of dog eat over the course of the month?

c. Which breed of dog ate the least amount of food during the month that Emmanuel graphed? How many pounds of food did this breed of dog eat over the course of the month?

d. How much more food did the breed that ate the most eat than the breed that ate the least?

e. How much more food did the beagle eat than the shih tzu?

7. Melissa found the following graph in a magazine at the library.

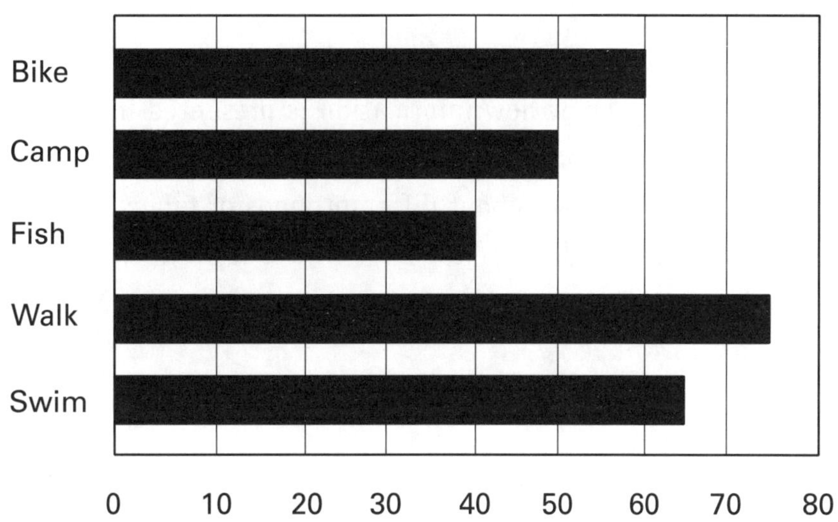

a. What does each bar represent?

b. Which activity has the most participants, according to the graph?

c. Which activity has the fewest participants?

d. How many more people say they walk than go fishing?

e. How many more people walk than bike?

Getting Information from Tables and Charts

LESSON SUMMARY

You already know how information is presented in graphs. Information can also be organized into tables and charts. In this lesson, you will learn about different kinds of tables and charts and how to read and understand the information in them.

Textbooks, almanacs, newspapers, websites, and yes, even math tests are filled with tables and charts. Tables and charts organize information. If you take just about any kind of standardized test, you are likely to be asked to find specific information in several different kinds of tables and charts.

GETTING INFORMATION FROM TABLES

Tables present information in rows and columns. Rows go across, or horizontally. Columns go up and down, or vertically. The box, or cell, that is made where a row and a column meet provides specific information. When looking for information in tables, it's important to read the table title, the column headings, and the row labels so you understand the information you are looking at. Let's look at some examples of tables and the types of information you might expect to learn from them.

Frequency tables are used to track how often things happen.

Example: Margie works in the school library. She keeps track of why students come to the library. Here is the frequency table she made for the students she spoke with last week.

Reasons for a Visit to the School Library ← Title

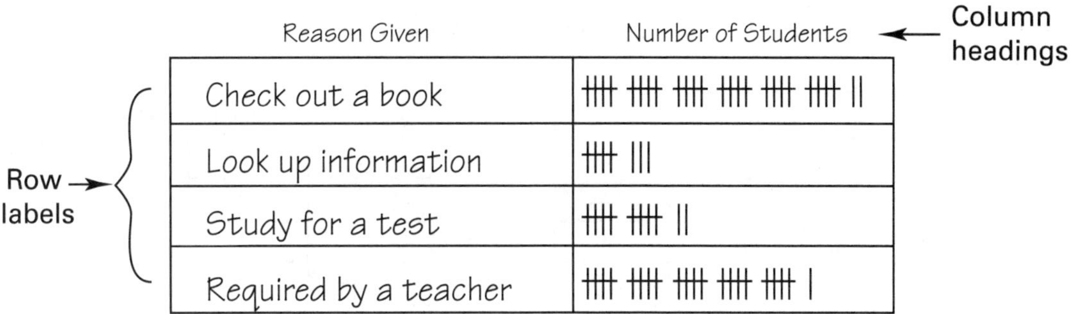

Reason Given	Number of Students
Check out a book	卌 卌 卌 卌 卌 卌 ||
Look up information	卌 |||
Study for a test	卌 卌 ||
Required by a teacher	卌 卌 卌 卌 卌 |

Based on Margie's frequency table, why did most of the students come to the school library last week?

Step 1: Look in the column containing tally marks. Each mark represents one student.

Step 2: Find the cell with the most tally marks in it.

If you look in the table, you will see that the first cell has the most tally marks in it.

Step 3: Trace your finger over to the reason given. This is the reason most of the students Margie tracked came to the school library last week. The reason most students visited the library last week was to check out a book.

Example: Look at Margie's frequency table in the last example. How many students came to the library to look up information last week?

Step 1: Find the row that says, "Look up information."

Step 2: Slide your finger over to the tally marks associated with this reason. Then, count the tally marks. Each mark represents one student, and each group of marks with a slash through it represents five students.

If you count the marks, you will see that there are eight (one group of five, plus three more marks). Therefore, eight students came to the library last week to look up information.

Example: Look at Margie's frequency table again. How many more students came to the library because their teacher told them to than the number who came to study for a test?

Step 1: Count the tally marks for *Required by teacher*.

There are 26 marks.

Step 2: Count the tally marks for *Study for a test*.

There are 12 marks.

Step 3: Subtract: 26 – 12 = 14. Therefore, 14 more students came to the library because they were required to by their teacher than those who came to study for a test.

Mileage tables are common in atlases and on maps. They tell you how many miles apart different places are.

Example: Maria's family used the following mileage table to plan their summer vacation this year. How far is it from Fairfield to Center?

Canyon Area Mileage Chart ← Title

Column headings →; Row labels ↓

	Lincoln	Fairfield	Plainview	Center	Smithville
Lincoln		65	48	80	98
Fairfield	65		108	67	145
Plainview	48	108		46	191
Center	80	67	46		92
Smithville	98	145	191	92	

Find Fairfield in the second row. Move your finger across until it's under the column heading *Center*. Or, you can begin by going to the column heading for Fairfield and move your finger down to the row label *Center*. The number in the cell represents the number of miles between the two cities. The answer is 67 miles.

No matter what kind of a table you are looking at, you can use the strategy in the last example to find the information requested. Read the column and row labels to locate the exact information requested.

Example: Look again at the mileage table Maria's family used. If her family decides to drive from Fairfield to Center, then from Center to Smithville, how many miles will they drive?

Step 1: Find the number of miles between Fairfield and Center. You know this already from the last example. It's 67 miles.

Step 2: Find the number of miles between Center and Smithville. It's 98 miles.

Step 3: Add the two numbers together: 67 + 92 = 159.

The answer is 159 miles.

PRACTICE

1. Use the mileage table below to answer the following questions. Answer the following questions. Then, check your answers at the end of the section.

U. S. CITY MILEAGE TABLE

	ATLANTA	BOSTON	CHICAGO	DENVER	LOS ANGELES
Atlanta		1,037	674	1,398	2,182
Boston	1,037		994	1,949	2,979
Chicago	674	994		996	2,054
Denver	1,398	1,949	996		1,059
Los Angeles	2,182	2,979	2,054	1,059	

a. How far is it from Atlanta to Boston?
b. Which is farther, Chicago to Denver or Boston to Chicago?
c. How many miles would you drive if you started in Denver, drove to Chicago, and then drove to Boston?
d. How much farther is Los Angeles to Boston than Atlanta to Chicago?

2. Use the table below to answer the following questions.

THE FUJITA-PEARSON TORNADO INTENSITY SCALE

CLASSIFICATION	WIND SPEED (IN MILES PER HOUR)	DAMAGE
F0	72	Mild
F1	73–112	Moderate
F2	113–157	Significant
F3	158–206	Severe
F4	207–260	Devastating
F5	261–319	Cataclysmic
F6	320–379	Overwhelming

a. If a tornado has a wind speed of 173 miles per hour, how would it be classified?
b. What kind of damage would you expect from a tornado having a wind speed of 300 miles per hour?
c. What wind speed would you anticipate if a tornado of F6 were reported?

3. Use the table below to answer the following questions.

COMPARTMENT 1	COMPARTMENT 2
Ax	Nozzle
Pry bar	Two lengths of hose
Sledgehammer	Pipe wrench
Torch	Toolbox
COMPARTMENT 3	**COMPARTMENT 4**
Fire extinguishers	First aid kit
Fire fighting foam	Oxygen cylinder
Portable pump	Rescue rope

a. What would be a good title for this table?
b. In which compartment would you expect to find more lengths of hose?
c. Which compartment would you look in for bandages?
d. Where would you find a sledgehammer?

4. Use the table below to answer the following questions.

CEDAR VALLEY SOFTBALL — TEAM A

NAME	AT BATS	HITS	WALKS
Meghan	16	5	2
Patty	17	6	3
David	15	7	2
Nasser	18	5	3
Lan	16	4	6

a. Which player has been up to bat the most?
b. Which player has had the most hits?
c. Either a hit or a walk can get a player to first base. Which player has most likely gotten to first base the most?
d. A player's batting average is the number of hits divided by the number of times at bat. What is Lan's batting average?

GETTING INFORMATION FROM CHARTS

Charts present information in many different ways. You probably use charts all the time. Here are three common kinds of charts.

FLOW CHART: to organize a series of steps in a process

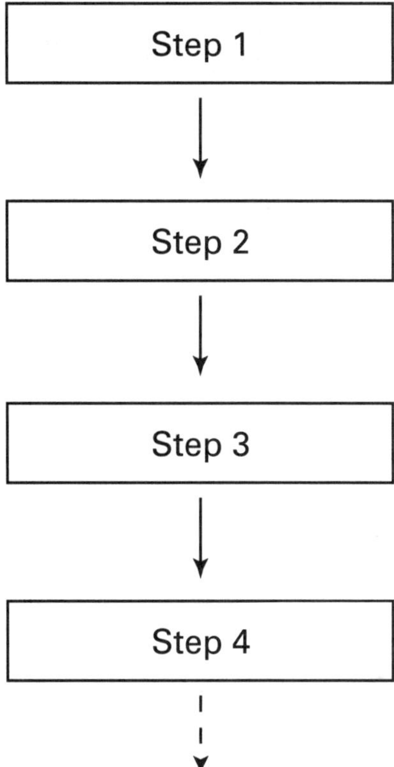

TIMELINE : to sequence events over time

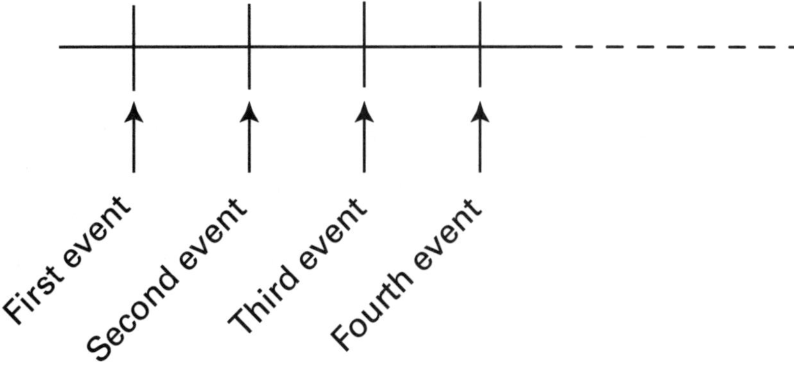

VENN DIAGRAMS:
to organize similarities and differences

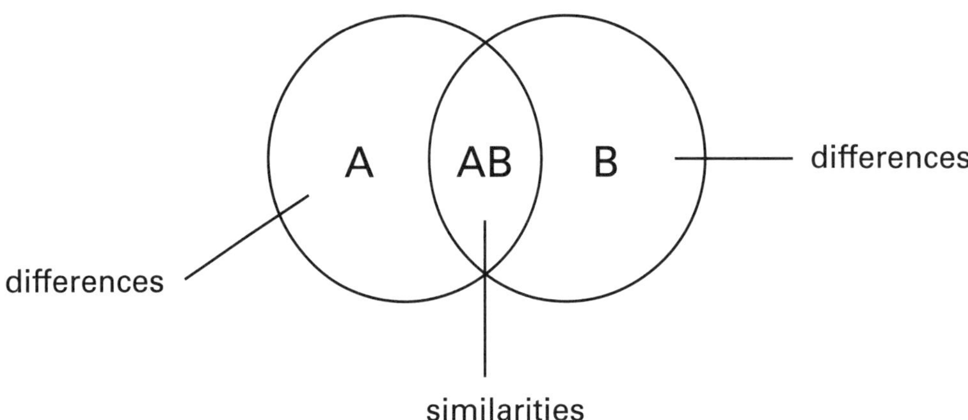

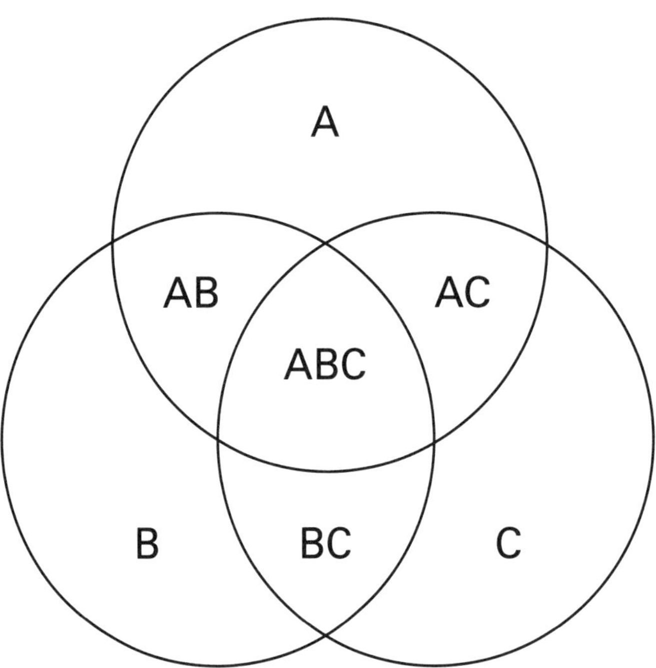

Let's look at some examples of these charts and the kinds of information you can get from each kind.

Flow Charts

Flow charts often show the steps in a process.

Example: Joachim is writing a report for school about corn. He found the following flow chart in a pamphlet from a corn processor.

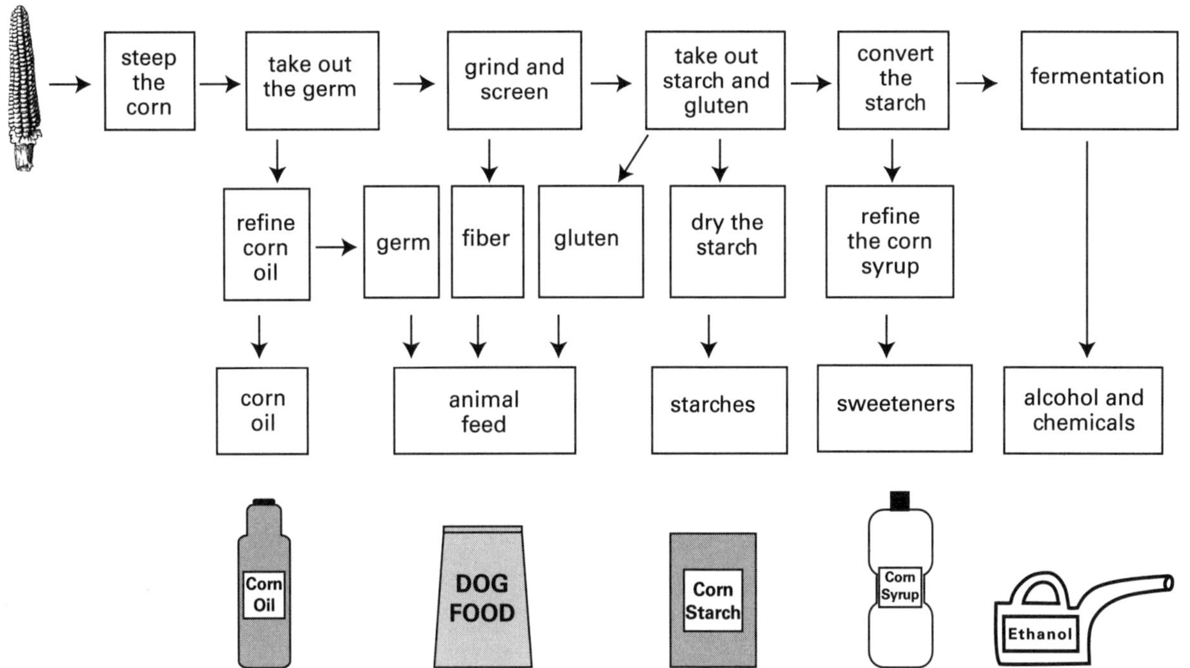

What five products are made using corn?

Look at the flow chart. At the bottom, the products of each process are listed. These are corn oil, animal feed, starches, sweeteners, and alcohol and other chemicals.

Example: Look at the flow chart. What three parts of corn are used to make food for animals?

Find animal feed in the chart. Then, follow the arrows to see which corn products are used to make it. These are germ, fiber, and gluten.

Example: Look at the flow chart again. Which is refined first, corn oil or corn syrup?

Look at the middle row of arrows in the flow chart. Notice that the corn oil is refined first. The corn syrup is refined last. So corn oil is refined before corn syrup.

Timelines

Timelines are used to show the sequence of events over time.

Example: Hakim found the following timeline in his history textbook.

Major Events of the Cold War
(1945 - 1955)

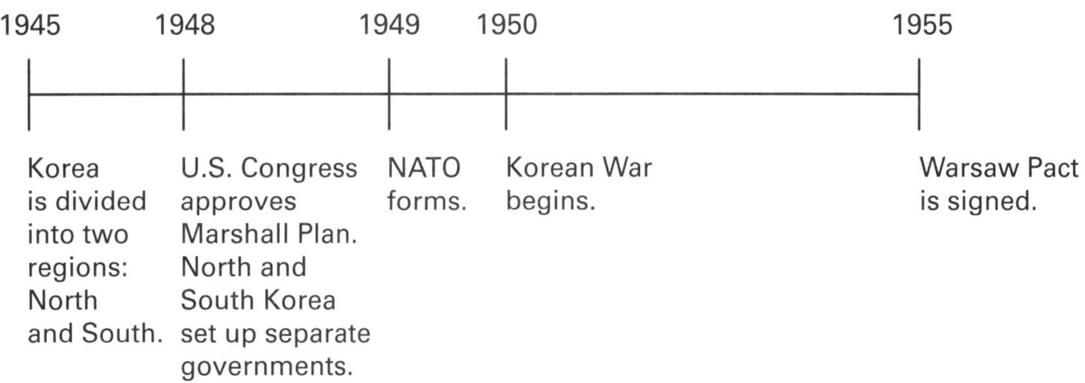

Did NATO form before or after the Korean War began?

Look at the dates on the timeline. Find when each event occurred. NATO formed in 1949. The Korean War began in 1950. So NATO formed before the Korean War began.

Example: Look at the timeline again. How many years passed between the division of Korea and the beginning of the Korean War?

Korea divided into two regions in 1945. The Korean War didn't begin until 1950. Subtract: 1950 – 1945 = 5. So five years passed between the division of Korea and the beginning of the Korean War.

Here are some things for you to keep in mind as you read a timeline.

- **Look at the time covered by the timeline. Ask yourself these questions: When does it begin? When does it end?**
- **Look at the sequence of events listed in the timeline. Ask yourself: How much time passes between each event listed in the timeline?**

continued from previous page

- **Think about events that are not listed in the timeline. Ask yourself: What other events were going on during this time?**
- **Think about how the events are related to one another. Ask yourself: Does one of these events cause another? How does each event relate to an earlier or a later event?**

Venn Diagrams

Venn diagrams are used to show how things are similar and different. They are often used in math when talking about sets of numbers. They can also be used to compare characteristics of groups.

Example: The following Venn diagram shows the kinds of socks and pants that students in Mrs. Scott's class were wearing on Monday.

Key:
S = Set of all students in Mrs. Scott's class on Monday
J = Set of students wearing jeans
C = Set of students wearing colored socks

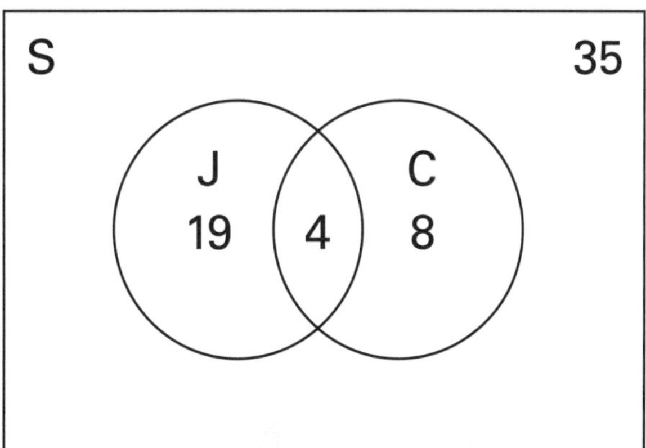

According to the diagram, how many students in Mrs. Scott's class were wearing jeans on Monday?

Look at the key. The circle labeled J represents all the students in Mrs. Scott's class who were wearing jeans on Monday. This circle contains the numbers 19 and 4. Add: $19 + 4 = 23$. So 23 students were wearing jeans in Mrs. Scott's class on Monday.

Example: Look at the Venn diagram again. How many students were wearing either jeans or colored socks but not both jeans and colored socks?

The key tells you that the circle labeled J represents all the students in Mrs. Scott's class who wore jeans on Monday. The circle labeled *C* represents all the students in Mrs. Scott's class who wore colored socks on Monday. The part of the two circles that overlaps represents students who were wearing both jeans *and* colored socks on Monday.

To answer this question, you need to add the number of students who wore either jeans or colored socks, but not both. So add 19 + 8 = 27. The answer is 27.

Example: Look at the Venn diagram again. How many students were in Mrs. Scott's class on Monday?

The box labeled *S* represents all the students present, or the total number of students, in Mrs. Scott's class on Monday. When people talk about Venn diagrams, they often call this box *The Universe*. It tells you what the largest set, or total, being discussed in the diagram is.

There were 35 students in Mrs. Scott's class on Monday.

Example: Look at the Venn diagram one last time. How many students were wearing neither jeans nor colored socks on Monday in Mrs. Scott's class?

To answer this question, you need to subtract the number of students wearing jeans and\or colored socks from the total number of students in class that day. You know there were 35 students in class. There were 31 (19 + 4 + 8) students wearing jeans and\or colored socks. Subtract: 35 – 31 = 4. There were only four students wearing neither jeans nor colored socks in Mrs. Scott's class on Monday.

PRACTICE

5. Malia is studying for a history exam. She made the following timeline to help her study. Answer the following questions using Malia's timeline. You can check your answers at the end of the lesson.

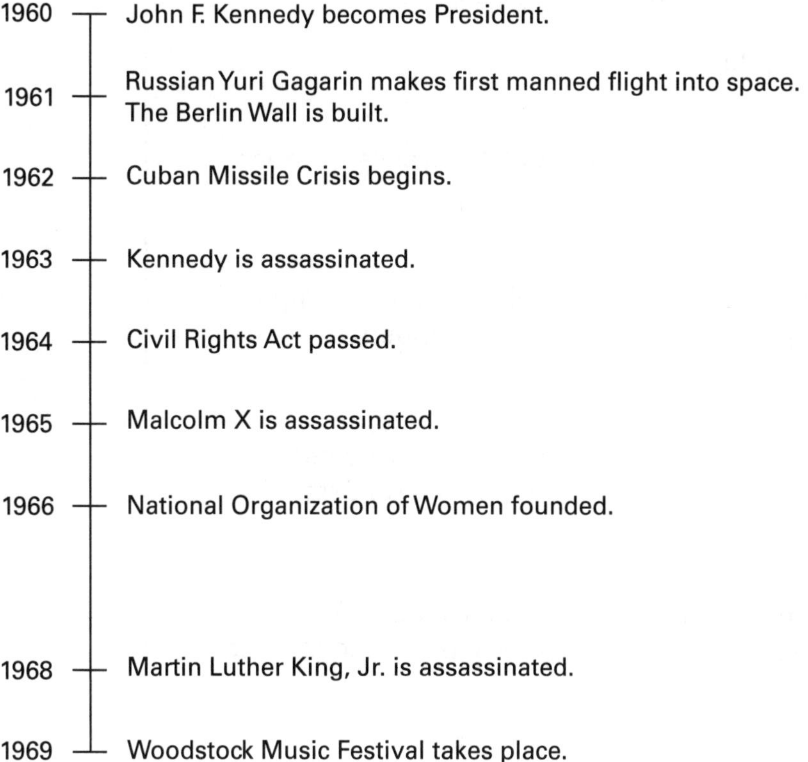

a. What period is covered by the timeline?
b. During what year was President Kennedy assassinated?
c. Who was assassinated first: President Kennedy, Martin Luther King, Jr., or Malcolm X?
d. Did Kennedy sign the Civil Rights Act?
e. Was the first person in space an American?

6. Masha is looking forward to getting her driver's license. She received the following flow chart in the mail from an insurance company. Answer the following questions using the flow chart. You can check your answers at the end of the lesson.

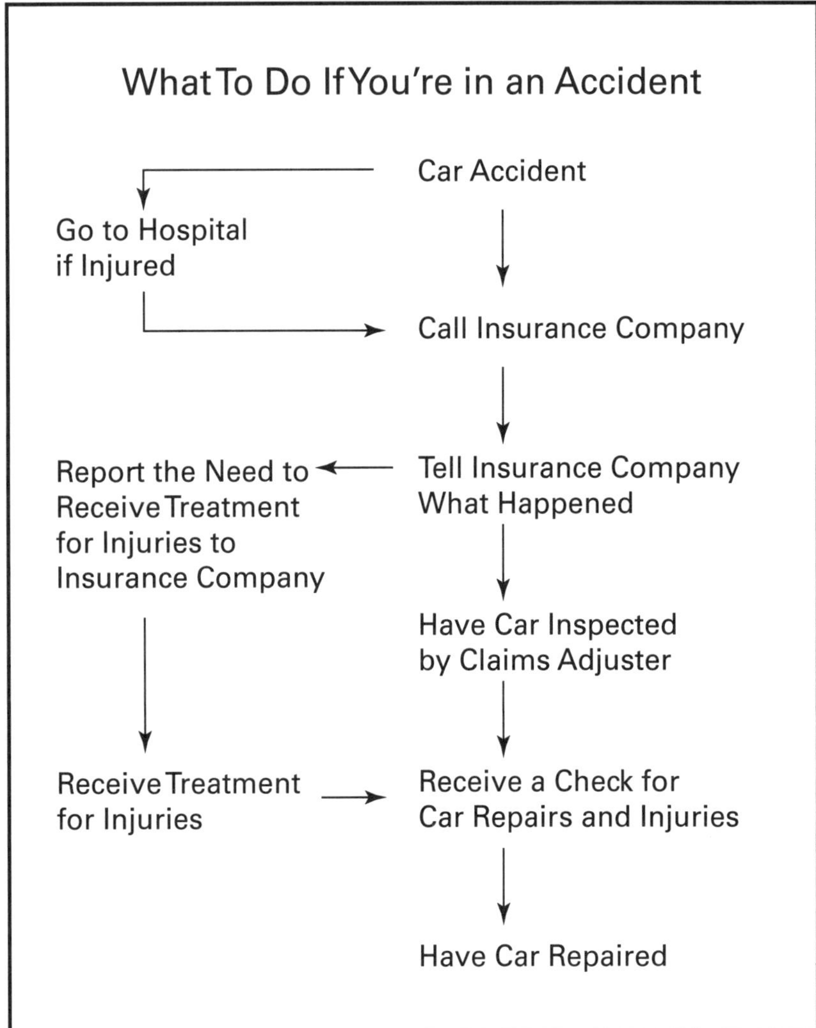

a. According to the flow chart, what is the first thing you should do if you are in a car accident?

b. When should you have your car repaired?

c. Which happens first: Your car is inspected by a claims adjuster or you receive a check for car repairs?

7. The U.S. Department of Agriculture puts out the following Food Guide Pyramid to help people choose a healthy diet. Use this chart to answer the following questions. You can check your answers that the end of the lesson.

The Food Guide Pyramid—A Guide to Daily Food Choices

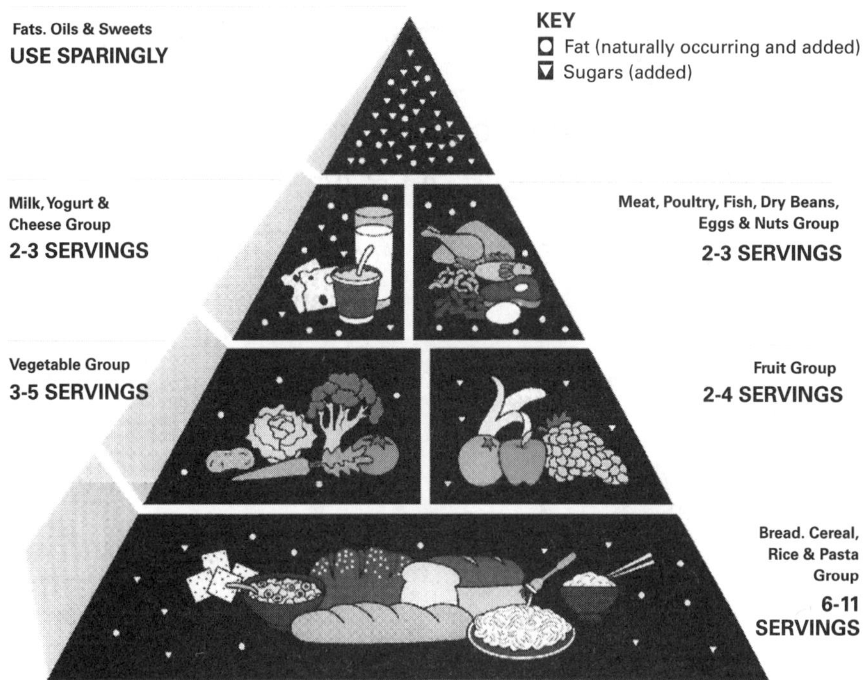

Source: U.S. Department of Agriculture/U.S. Department of Health and Human Services

a. How many servings of vegetables are recommended each day?
b. Which food group should you eat the most servings of each day?
c. Based on the key, which group of foods has the most added fats and sugars?

8. Morrison Middle School had the following basketball scores last month:

101	89	112	108
82	104	88	120
111	109	85	91
122	93		

Draw a Venn diagram similar to the one shown below. Then write the scores where they belong in the diagram. You can check your answers at the end of the lesson.

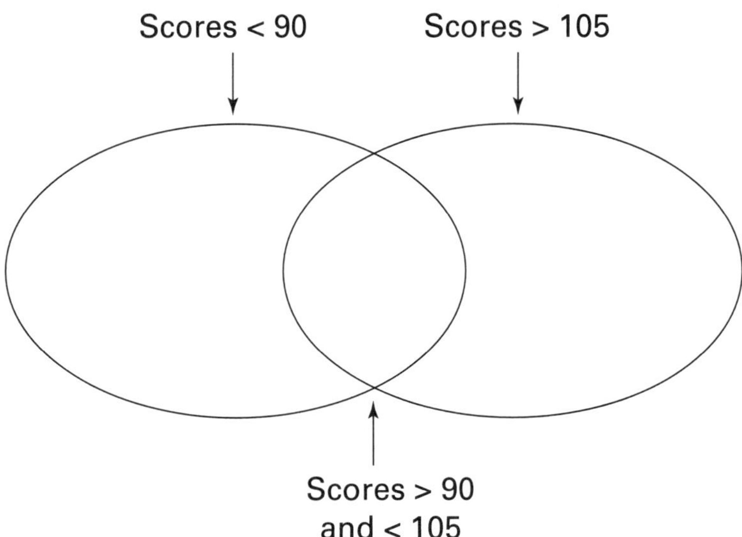

Real World Problems

These problems apply the skills you've learned in Section 6 to everyday situations. As you work through these problems, you'll see that the skills you've learned in this section aren't only important for math tests. They are important skills for ordinary questions that come up every day. You can check your answers at the end of the lesson.

Look at the labels on tables and charts to help you find the information you are looking for.

1. This summer, Tennille operates a drink stand at the zoo. The graph below shows the types of drinks she sells on a typical day.

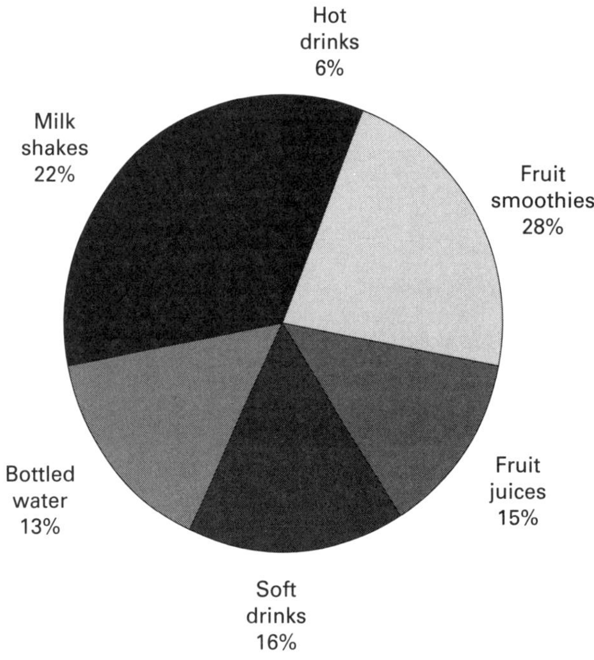

a. If Tennille sells 325 drinks one day, how many fruit smoothies is she most likely to sell?
b. What is the most popular kind of drink Tennille sells?
c. What is the least popular kind of drink she sells?

2. The table below lists the fires in the Tri-County Region for the month of June.

FOREST FIRES, TRI-COUNTY REGION

JUNE 2002

DATE	AREA	NUMBER OF ACRES BURNED	PROBABLE CAUSE
June 2	Burgaw Grove	115	Lightning
June 3	Fenner Forest	200	Campfire
June 7	Voorhees Air Base Training Site	400	Equipment Use
June 12	Murphy County Nature Reserve	495	Children
June 13	Knoblock Mountain	200	Miscellaneous
June 14	Cougar Run Ski Center	160	Unknown
June 17	Fenner Forest	120	Campfire
June 19	Stone River State Park	526	Arson
June 21	Burgaw Grove	499	Smoking
June 25	Bramley Acres Resort	1,200	Arson
June 28	Hanesboro Crossing	320	Lightning
June 30	Stone River State Park	167	Campfire

a. What period does the table cover?
b. Where did a fire occur one week after the Voorhees Air Base fire?
c. Which fires were probably caused by arson?
d. Which fire burned the most acres?

3. Ivan found the following information in a pamphlet from his local waste disposal center.

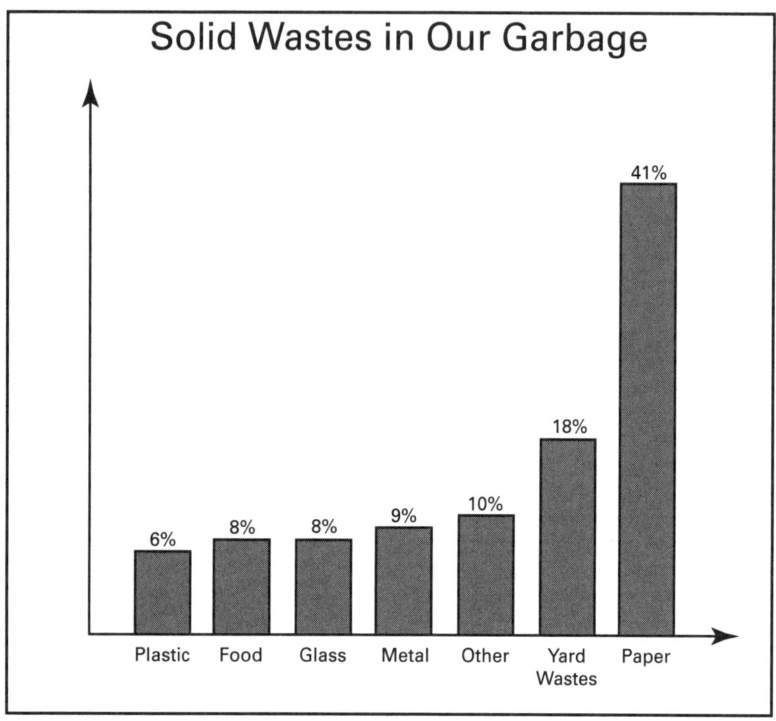

WHERE OUR GARBAGE GOES

METHOD OF DISPOSAL	PERCENTAGE OF GARBAGE
Landfill	80
Recycled	11
Converted to energy	6
Burned	3

a. What kind of solid waste makes up the largest portion of garbage?
b. What percent of solid waste comes from yards?
c. Where does most of our garbage go?
d. If you were to go to a landfill, what kind of garbage would you probably see the most of?
e. What percent of garbage is being recycled?

4. Use the table to answer the following questions.

HURST COUNTY TOWNS, NUMBER OF DAYS WITHOUT SIGNIFICANT PRECIPITATION*

TOWN	NUMBER OF DAYS	STATUS**
Riderville	38	level two
Adams	25	level one
Parkston	74	level three
Kings Hill	28	level two
West Granville	50	level three
Braxton	23	level three
Chase Crossing	53	level four
Livingston Center	45	level three

**At least half an inch in a 48-hour period.*

***The higher the level, the greater potential for fire.*

a. What is the status of the town with the fewest number of days without significant precipitation?
b. Which town has the most days without significant precipitation?
c. Which towns are Status Three?
d. How many more days without significant precipitation does the town with the most days have than the town with the least days?

5. Marissa's mother received the following information along with her water bill this month.

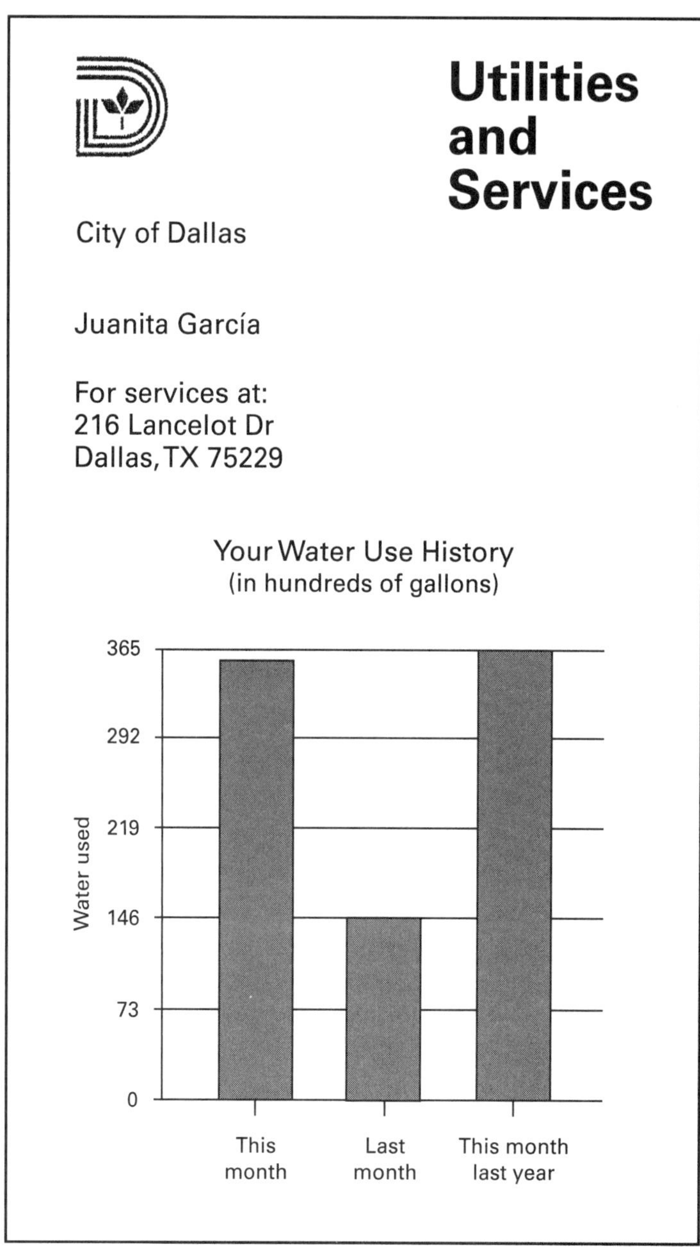

a. During which month shown in the graph did the Garcia family use the most water?

b. How many gallons of water did the Garcia family use this month last year?

c. How much more water did the Garcias use this month last year than they did last month?

d. If the Garcias used 35,700 gallons of water this month and paid $74.00 for water, how much were they charged per gallon?

6. Alex's little brother went to the doctor for his two-year check-up. Every time he goes to the doctor, the doctor checks the toddler's weight and height against the following growth charts.

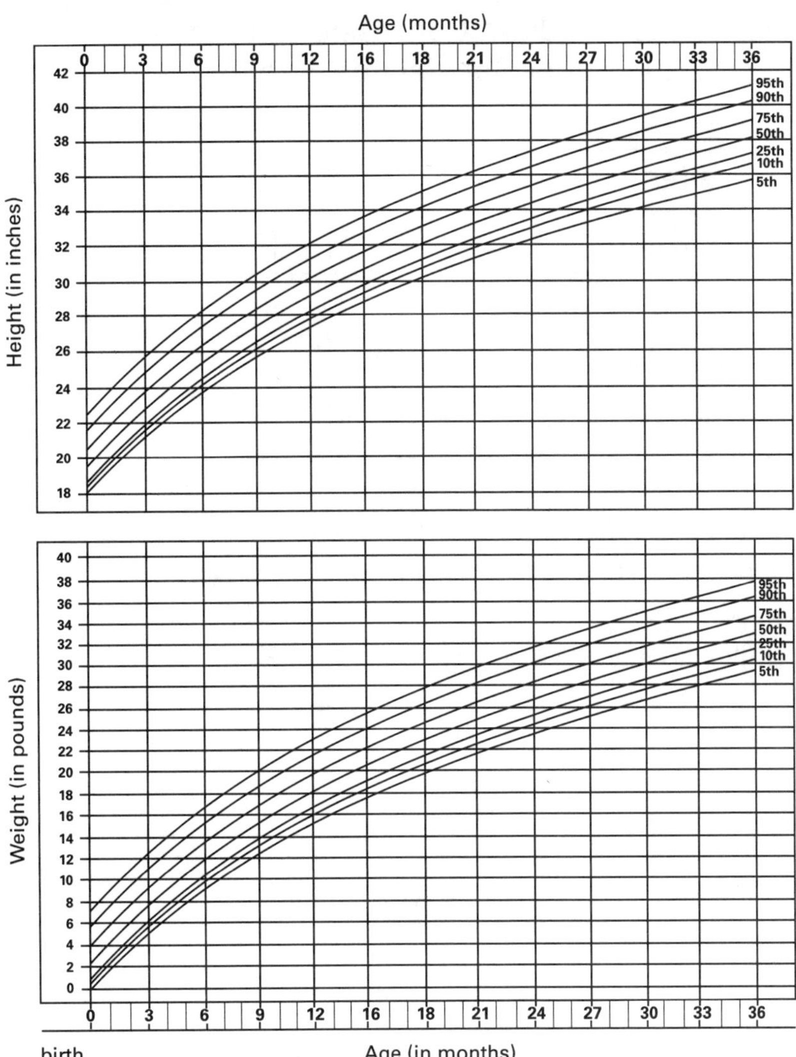

a. His little brother weighed 28 pounds and was 34 inches tall. What percentile was his little brother in for weight and height?

b. A child in the 95th percentile is bigger than about 95% of other children the same age and sex. How tall is a 33-month-old child in the 95th percentile?

c. Alex's little brother weighed 6 pounds at birth. What percentile was he in?

7. Connie worked at the football concession stand this season. She randomly asked customers to fill out a card about the concession stand's service. Then, she tallied the responses. Below is Connie's frequency table after one football game.

Tiger Stadium Concession Stand
Customer Satisfaction Survey

Excellent	𝍸 𝍸 𝍸
Very Good	𝍸 𝍸 ||
Average	𝍸 𝍸
Below Average	𝍸 |||
Poor	𝍸 𝍸 𝍸

a. How many people said that the concession stand did an excellent or very good job?
b. What is the ratio of customers who said the service was excellent to those who said the service was poor?
c. What percent of customers said the service was poor?

SECTION

6

Answers & Explanations

LESSON 14

1. a. Oxygen makes up 46% of the Earth's crust.
b. Oxygen makes up 46%. Silicon makes up 28%. Together, these two elements make up 74% (28% + 46%) of the Earth's crust.
c. Iron makes up 6%. Aluminum makes up 8%. Together, these two elements make up 14% (6% + 8%) of the Earth's crust.

2. a. Energy: voters say they'd like about 10% of the budget spent on energy and about 11% is spent on energy.
b. Health
c. National defense
d. Voters said they wanted money to go to health and environment. Defense and health received the most money.

3. a. Housing; 30%
b. Clothing; 4%
c. According to the graph, 22% of the budget is spent on food. Transportation and clothing combined make up 13% (9% + 4%) of the budget. So, 22 – 13 = 9. Thus, 9% more money is spent on food than on transportation and clothing combined.
d. According to the graph, 10% is spent on savings. Ten percent of $2,000 is $200 (0.1 × 2,000). So, the Johnsons save $200 each month.

4. a. Look at the labels. The plant's height is measured in inches on the vertical axis. Time in days is measured on the horizontal axis.

b. Use a ruler or a straight edge of paper to guide your eye up from day 9 on the graph. Then trace over to the height. The plant was probably about 4 inches tall on day 9.

c. Look at how steep the line is. The steepest part of the line represents the fastest growth. The line is the steepest between days 19 and 21.

d. Look at how steep the line is. The flattest part of the line represents the slowest growth. The line is flattest between days 23 and 25.

e. The plant was about 1 inch tall on day 1. It was about 10 inches tall on day 25. Subtract: $10 - 1 = 9$. The plant grew about 9 inches over the time that Edwin measured it.

5 a. Look at the labels. The percent of workers using each form of transportation is shown on the vertical axis. Population density in workers per acre is shown on the horizontal axis.

b. First, find the line that represents "using public transportation to get to work." Where is this line highest on the graph? When the population density is highest, at 150 workers per acre, 70% of workers use public transportation in the more populated areas, more than at any other density. Therefore, people are more likely to use public transportation in areas where the population is most dense: 150 workers per acre.

c. First, find the line that represents "using your own car to get to work." Look for the point where this line is highest on the graph. When the population is least dense, the line is at 100%, higher than it is at any other point on the graph. Therefore, people are most likely to drive their own car to work in areas where the population is not very dense: 1 worker per acre.

d. Find the line that moves down as population density increases. It's the line representing "using your own car to get to work." This is the form of transportation that decreases as population density increases.

6. a. Read the labels. Each bar represents the amount of dog food in pounds a different breed of dog consumed in one month.

b. The tallest bar represents the most food eaten. The mastiff ate about 42 pounds of food—more than any other breed shown.

c. The shortest bar represents the smallest amount of food eaten. The shih tzu ate about 4 pounds of food—less than any other breed shown.

d. Subtract: $42 - 4 = 38$. The mastiff ate 38 more pounds of food than did the shih tzu.

e. Subtract the amount the shih tzu ate from the amount the beagle ate: $16 - 4 = 12$. The beagle ate 12 more pounds of food than did the shih tzu.

7. a. Each bar represents the number of millions of people who participate in a given sport.

b. The longest bar represents the most popular sport: walking.

c. The shortest bar represents the sport shown with the least number of participants: fishing.

d. About 75 million people say they walk. About 40 million say they fish. Subtract: $75 - 40 = 35$. So about 35 million more people walk than fish.

e. About 75 million people say they walk. About 60 million say they bike. Subtract: 75 – 60 = 15. So about 15 million more people walk than bike.

LESSON 15

1. a. 1037 miles
   b. Chicago to Denver is 996 miles. Boston to Chicago is 994 miles. So Chicago to Denver is farther than Boston to Chicago.
   c. Denver to Chicago is 996 miles. Chicago to Boston is 994 miles. Add the two: 994 + 996 = 1,990. So you would drive 1,990 miles.
   d. Los Angeles to Boston is 2,979 miles. Atlanta to Chicago is 674 miles. Subtract: 2,979 – 674 = 2,305. The answer is 2,305.

2. a. F3
   b. cataclysmic
   c. 320 – 379 miles per hour

3. a. Answers will vary. One possible title is "Fire Fighting Tools"
   b. Compartment 2
   c. Look in the first aid kit in Compartment 4.
   d. Compartment 1

4. a. Nasser—18 times up to bat
   b. David—7 hits
   c. Add the number of hits and walks for each player.
      Meghan: 5 + 2 = 7
      Patty: 6 + 3 = 9
      David: 7 + 2 = 9
      Nasser: 5 + 3= 8
      Lan: 4 + 6 = 10
      It is most likely that Lan has reached first base the most.
   d. Divide: 4 ÷ 16 = .25 So, Lan's batting average is .250 (you add a zero because there are always three decimal places in a batting average).

5. a. 1960 – 1969
   b. 1963
   c. Kennedy was assassinated in 1963; Malcolm X was assassinated in 1965; Martin Luther King, Jr., was assassinated in 1968. So the answer is Kennedy.
   d. No, Kennedy could not have signed the Civil Rights Act because it was passed a year after his death.
   e. No, the timeline says that the first person in space was a Russian.

6. **a.** If you are injured, you first go to the hospital. If you're not injured, you call the insurance company.
b. The car is repaired last—after you receive a check for the repairs from the insurance company.
c. First, your car is inspected. Then, you receive a check for the repairs.

7. **a.** 3–5 servings per day
b. Bread, cereal, rice, and pasta
c. Fats, oils, and sweets have the most added fats and sugars.

8.

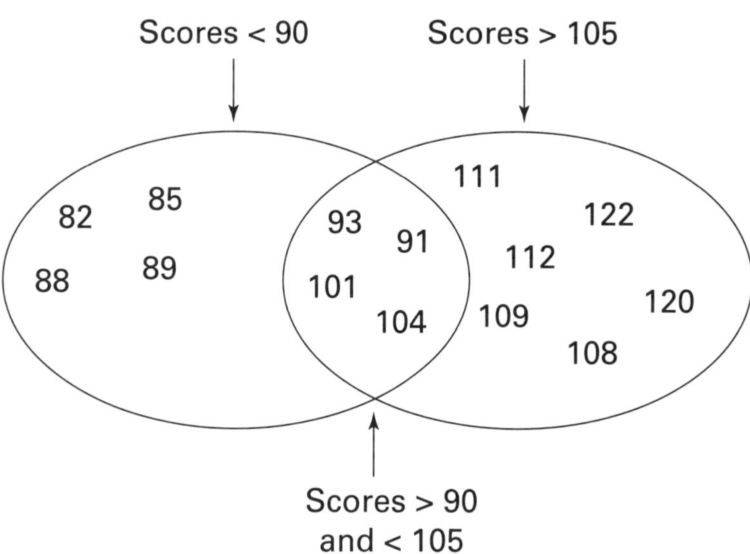

▶ REAL WORLD PROBLEMS

1. **a.** Fruit smoothies make up 28% of her sales. Multiply: $0.28 \times 325 = 91$. If Tennille sells 325 drinks in one day, she most likely sells about 91 fruit smoothies.
b. The largest slice of the pie chart is 28%—fruit smoothies. Fruit smoothies are the most popular drinks Tennille sells.
c. Find the smallest slice of the pie chart. It's 6%—for hot drinks. Hot drinks are the least popular kind of drinks Tennille sells.

2. **a.** June 2002
b. Cougar Run Ski Center
c. Stone River State Park and Bramley Acres Resort
d. Bramley Acres Resort

3. **a.** Look at the bar graph: paper—41%.
b. Look at the bar graph: 18%.
c. Look at the table: landfill.
d. From the bar graph, you know that 41% of solid waste is paper. Most of this goes to landfills. So if you visit a landfill, you should find a lot of paper.
e. Look at the table: 11%.

4. **a.** Braxton has the fewest number of days without significant precipitation: 23 days. It is level three.
b. Parkston has the most days without significant precipitation: 74.
c. Parkston, West Granville, Braxton, and Livingston Center are all level three.
d. Parkston has the most days without significant precipitation: 74 days. Braxton has the fewest number of days without significant precipitation: 23 days. Subtract: $74 - 23 = 51$ days. The answer is 51 days.

5. **a.** The tallest bar represents the most water usage. The Garcia family used the most water this month last year.
b. The bar representing water usage for this month last year is 365. The label says that the numbers represent hundreds of gallons. So, this month last year, the Garcias used $365 \times 100 = 36{,}500$. The answer is 36,500 gallons.
c. Find the amounts of water used this month last year and last month. Then subtract: $36{,}500 - 14{,}600 = 21{,}900$. The answer is 21,900 gallons.
d. You want to know the price per gallon, so divide $74.00 \div 35{,}700 = 0.002$. The answer is less than one penny!

6. **a.** First, find Alex's little brother's age: 24 months. Then move up to find his weight: 28 pounds. Trace the line that his weight falls on to the percentile label: Alex is in the 50th percentile for weight. (This means that about half of two-year-old boys weigh more and about half weigh less than Alex's little brother.) Now do the same things for Alex's little brother's height. He's in the 25th percentile for height. (This means that about 75% of boys are taller than Alex's little brother.)
b. First, find 33 months on the height chart. Move your finger down to the 95th percentile line. The point where these intersect is the height of a 33-month-old child in the 95th percentile: 40 inches.
c. Find 6 pounds and birth on the weight chart. Then follow the percentile line up to its label. Alex's little brother was in the 5th percentile at birth for weight. (That means that most other babies weighed more than he did at birth.)

7. **a.** Count the tally marks beside "Excellent" and "Very good": There are 27. The answer is 27 people.

b. Count the number of marks beside each category. There are 15 marks beside "Excellent" and 15 marks beside "Poor." So the ratio is 15:15, which reduces to 1:1. The answer is 1:1, or $\frac{1}{1}$.

c. Count the number of customers who said the service was poor: 15. Then count the total number of customers surveyed: 60. Write a fraction using these two numbers: $\frac{15}{60}$, which reduces to $\frac{1}{4}$. The answer is 25%.

SECTION

7

Introduction to Algebra

ALGEBRA IS THE branch of mathematics that denotes quantities with letters and uses negative numbers as well as ordinary numbers. There are many faces of algebra. You can interpret a sentence and translate it into algebra. You can take an algebraic expression and turn it into words. You may work with a "solve for *x*" question, or a "simplify the equation" question. This section will give you a basic introduction to algebra, so you understand how you can use it to solve problems in the real world.

Algebraic Expressions

LESSON SUMMARY

This lesson will introduce you to the basic vocabulary of algebra. You will learn what an algebraic expression is. You will also learn how to evaluate and simplify algebraic expressions. In short, this lesson will prepare you for the next lesson on solving algebraic equations.

Algebra is a branch of math that uses letters to represent unknown numbers. A letter that represents an unknown number is called a *variable.* You often use algebra to translate everyday situations into a math sentence so that you can then solve problems.

WHAT ARE ALGEBRAIC EXPRESSIONS?

An *algebraic expression* is a group of numbers, variables (letters), and operation signs (+, –, ÷, and so on). Variables are usually written in italics. For example, the x in the following algebraic expressions is the variable:

$$5x + 2 \qquad 3x - 8 \qquad \frac{4x}{9}$$

Any letter can be used to represent a number in an algebraic expression. The letters, x, y, and z are commonly used.

Algebraic expressions translate the relationship between numbers into math symbols. The following are some examples.

MATH RELATIONSHIP IN WORDS	TRANSLATED INTO AN ALGEBRAIC EXPRESSION
A number plus six	$x + 6$
Five times a number	$5x$
Three less than a number	$x - 3$
The product of seven and a number	$7x$
A number divided by eight	$x \div 8$ or $\frac{x}{8}$
A number squared	x^2

When multiplying a number and a variable, you just have to write them side by side. You don't need to use a multiplication symbol.

Notice that the value of an algebraic expression changes as the value of the variable (here, x) changes. Let's look at the first algebraic expression listed above: $x + 6$.

If $x = 1$, then $x + 6 = 1 + 6 = 7$.
If $x = 2$, then $x + 6 = 2 + 6 = 8$.
If $x = 3$, then $x + 6 = 3 + 6 = 9$.
If $x = 100$, then $x + 6 = 100 + 6 = 106$.
If $x = 213$, then $x + 6 = 213 + 6 = 219$.

Example: Translate the following math relationship into an algebraic expression: the quotient of three divided by a number plus the difference between three and two.

Step 1: You know that the word *quotient* means to divide, so you write

$3 \div$

Step 2: The word *number* represents your variable, so you write

$3 \div x$

Step 3: The word *plus* means to add, so you write

$3 \div x +$

Step 4: The word *difference* means to subtract, so you write

$3 \div x + (3 - 2)$

Remember from Lesson 1 (the order of operations), that parentheses are used to indicate an operation that should be performed first.

So, the answer is $3 \div x + (3 - 2)$.

When translating a math relationship into an algebraic expression, keep these key words in mind.

TRANSLATING MATH RELATIONSHIPS INTO ALGEBRAIC EXPRESSIONS

THESE WORDS	OFTEN TRANSLATE INTO THESE MATH SYMBOLS
added to sum plus increased by combine altogether	+
subtracted from difference decreased by minus take away less	–
multiplied by product times	$\times$, • ( ) Parentheses can also indicate multiplication. 3(5) is the same as 3×5 or 3 • 5
divided by quotient per	÷
equals is are	=
a number	x

Remember that when dividing and subtracting the order of the numbers is very important. Here are some examples.

- **The difference between three and two means 3 – 2 (which equals 1).**
- **Three less than two means 2 – 3 (which equals –1).**
- **The quotient of four divided by two means 4 ÷ 2 (which equals 2).**

It's the same when variables are used.

- **The difference between three and a number means 3 – x.**
- **Three less than a number means x – 3.**
- **The quotient of four and a number means 4 ÷ x**
- **The quotient of a number divided by four means x ÷ 4.**

Example: Translate the following math relationship into an algebraic expression: one-fourth a number is decreased by nine.

Step 1: You know that the word *one-fourth* is a fraction, so you write

$\frac{1}{4}$

Step 2: The word *number* represents your variable, and when you are talking about fractions, the word *of* indicates that you should multiply, so you write

$\frac{1}{4}x$

Step 3: The words *decreased by* mean to subtract, so you write

$\frac{1}{4}x - 9$

Thus, the answer is $\frac{1}{4}x - 9$.

PRACTICE

Translate each math relationship into an algebraic expression. You can check your answers at the end of the section.

1. ten minus twice a number
2. the difference between six and a number
3. eleven less than the sum of five and a number
4. seven times a number plus three
5. four times a number increased by two
6. the quotient of a number divided by two
7. eight times a number minus one
8. ten minus the sum of fourteen and a number
9. the product of three and x and y
10. five divided by the product of x and y

▶ EVALUATING ALGEBRAIC EXPRESSIONS

Algebraic expressions have specific values only when the variables have values. Finding the value of an algebraic expression by plugging in the known values of its variables is called *evaluating* an expression.

Example: Evaluate $5 \div (x + y)$, when $x = 2$ and $y = 3$.

Step 1: Plug the numbers into the algebraic expression.

$5 \div (2 + 3)$

Step 2: Solve by following the order of operations.

Add the numbers in parenthesis: $5 \div (2 + 3) = 5 \div 5$

Divide: $5 \div 5 = 1$

The answer is 1.

Whether you are working with numbers or variables, the order of operations is the same. Remember the order of operations from Lesson 1—Please Excuse My Dear Aunt Sally. (Parentheses, Exponents, Multiplication, Division, Addition, Subtraction)

THE ORDER OF OPERATIONS	
STEP 1:	Do all the operations in parentheses. Then solve for exponents.
STEP 2:	Multiply and divide numbers in order from left to right.
STEP 3:	Add and subtract numbers in order from left to right.

You might want to review Lesson 1, if this isn't familiar to you.

Example: Find the value of $5 + x$, when $x = -2$.

Step 1: Plug the numbers into the algebraic expression.

$5 + (-2)$

Step 2: Solve.

$5 - 2 = 3$

Thus, the answer is 3.

Variables can represent any quantity. A variable might be a whole number, a decimal, or a fraction, for example. A variable can also represent a *signed number*—a number with a positive (+) or a negative (–) sign in front of it. Any number can have a sign in front of it. If a number has no sign in front of it, it is + (positive).

Signed whole numbers and zero make up a group of numbers called *integers*. Integers are often represented on a *number line* like the one shown below. Zero (0) is in the center of the line, and numbers to the left of the zero are negative, while the numbers to the right of zero are positive.

You can use a number line to add and subtract signed whole numbers. Here's how it works.

EXAMPLE: 8 + (–4)

Step 1: 8 is positive since there is no negative sign in front of it.

Step 2: Start at zero. Then, move 8 units in the positive direction (to the right), as shown below.

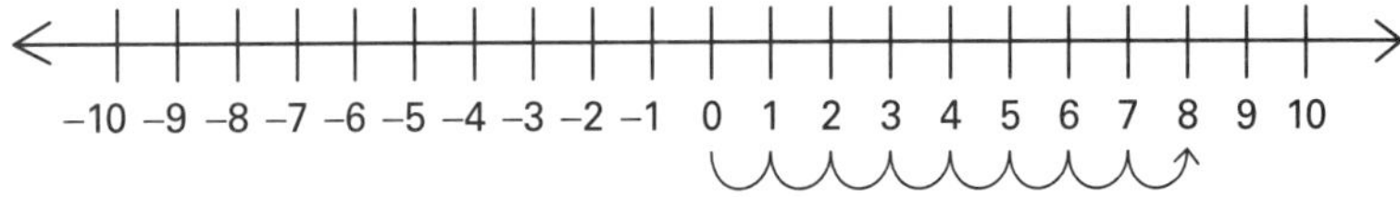

Step 3: The sign in front of the 4 is – (negative). So you need to move 4 units in the negative direction (to the left) as shown below.

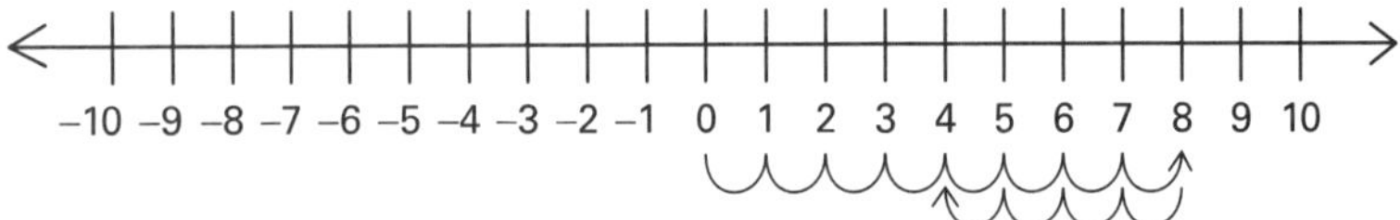

Thus, the answer is 4.

On the next page are some basic rules to help you add, subtract, multiply, and divide signed numbers.

continued from previous page

WHAT SIGN IS THE ANSWER?

RULE	EXAMPLE
When adding	
If the numbers have the same sign, just add them together. The answer has the same sign as the numbers being added.	$5 + 6 = +11$ $-5 + (-6) = -11$ $1 + 2 + 3 = +6$ $-1 + (-2) + (-3) = -6$
If two numbers have different signs, subtract the smaller number from the larger one. The answer has the same sign as the larger of the two numbers.	$5 + (-6) = -1$ $-5 + 6 = +1$
If you are adding more than two numbers, add the positive numbers and the negative numbers separately. Then, follow the rule above.	$+2 + (-5) + (-7) + 4 + 3 =$ $2 + 4 + 3 = +9$ $(-5) + (-7) = -12$ $+9 + (-12) = -3$
When subtracting	
Change the sign of the number that follows the minus sign. Then, add. (Notice that two negative signs next to each other make a positive sign.)	$3 - 5 = 3 + (-5) = -2$ $-3 - 5 = -3 + (-5) = -8$ $-3 - (-5) = -3 + (+5) = +8$
When multiplying	
If you are multiplying two numbers and the numbers have the same sign, then the answer is positive. (This also applies to any even number of numbers being multiplied together.)	$2 \times 6 = 12$ $-2 \times (-6) = 12$
If you are multiplying two numbers and the numbers have the opposite signs, then the answer is negative.	$-2 \times 6 = -12$ $2 \times (-6) = -12$
When multiplying more than two numbers together, two negative signs make a positive sign.	$-2 \times (-6) = 12$ $-1 \times (-2) \times (-3) = -6$ $-1 \times (-2) \times (-3) \times (-4) = 24$
When dividing	
If the numbers have the same sign, then the answer is positive.	$15 \div 3 = 5$ $-15 \div (-3) = 5$
If the numbers have the different signs, then the answer is negative.	$-15 \div 3 = -5$ $15 \div (-3) = -5$

PRACTICE

Evaluate each algebraic expression using the numbers given. You can check your answers at the end of the section.

11. $x + 2x + 10$
   a. when $x = 1$
   b. when $x = -1$
   c. when $x = 5$

12. $x^2 + 10$
   a. when $x = 1$
   b. when $x = -1$
   c. when $x = 3$

13. $2(x - 1)$
   a. when $x = -1$
   b. when $x = -2$
   c. when $x = 3$

14. $5(x - 3) + 2y$
   a. when $x = -1$ and $y = 1$
   b. when $x = 3$ and $y = -2$
   c. when $x = 5$ and $y = -1$

15. $3x + \frac{2y}{3}$
   a. when $x = -1$ and $y = 1$
   b. when $x = 5$ and $y = -3$
   c. when $x = 12$ and $y = -12$

Notice where a negative sign is in an algebraic expression. It can make a difference if the negative sign is inside parentheses or outside them. Compare these expressions.

$$(-2)^2 = (-2)(-2) = 4$$
$$-(2)^2 = -(2)(2) = -4$$

SIMPLIFYING ALGEBRAIC EXPRESSIONS

The parts of an algebraic expression are called *terms*. A term is a number or a number and the variables associated with it. For example, the following algebraic expression has three terms:

Algebraic expression: $5x^2 - 5x + 1$
Terms: $+5x^2$, $-5x$, $+1$

As you can see, the terms in an algebraic expression are separated by + and – signs. Notice that the sign in front of each term is included as part of that term—the sign is *always* part of the term. If no sign is given, then the term is positive (+).

Still not sure how to break an algebraic expression into its terms? Try these. Break each algebraic expression into its terms.

1. $2x^2 - 6x - 5$
2. $3xy + x^2$
3. $5xyz - 17$
4. $4x^2 - 3x - 6y^2$
5. $12 - 20x^2$

Answers:

1. Three terms: $2x^2$, $-6x$, -5
2. Two terms: $3xy$, x^2
3. Two terms: $5xyz$, -17
4. Three terms: $4x^2$, $3x$, $6y^2$
5. Two terms: 12, $-20x^2$

Simplifying an algebraic expression means to combine like terms. *Like terms* are terms that use the same variable and are raised to the same power. Here are some examples of like terms.

LIKE TERMS	BOTH TERMS CONTAIN
$5x$ and $9x$	x
$3x^2$ and $8x^2$	x^2
xy and $-5xy$	xy

When you combine like terms, you group all the terms that are alike together. This makes it easier to evaluate the expression later on.

Example: Simplify the following algebraic expression: $5x + 3y + 9x$.

Step 1: Write the like terms next to each other.

You know that $5x$ and $9x$ are like terms because they both have x. So write them next to each other: $5x + 9x + 3y$

Step 2: Combine the like terms.

Add the like terms:

$5x + 9x + 3y = 14x + 3y$

You can't simplify the $3y$ because there are no other terms in the expression with the variable y. So you leave it alone.

The simplified expression is $14x + 3y$.

Example: Simplify the following algebraic expression: $2x^2 + 3y + 9xy + 3y^2$

Step 1: Write the like terms next to each other.

There are no like terms in this expression. It's already simplified.

Example: Simplify the following algebraic expression: $5x(2x - 1) + 9x$

Step 1: Write the like terms next to each other. Begin by distributing the $5x$ so that you can remove the parentheses.

Multiply: $5x(2x - 1)$

$10x^2 - 5x$

So now your equation is $10x^2 - 5x + 9x$

Now you can see that $-5x$ and $9x$ are like terms because each contains the x term.

Step 2: Combine the like terms.

Add the like terms: $-5x + 9x = 4x$

Now your equation is simplified to $10x^2 - 4x$

You can't simplify further because there are no other terms in the expression with the same variable. Therefore, the simplified expression is $10x^2 - 4x$.

When combining like terms, begin by solving to remove the parentheses. If a negative sign comes in front of a set of parentheses, it affects *every* term inside the parentheses. Here's an example.

EXAMPLE: $-2(x - 3y + 2)$

When you multiply each term inside the parentheses by -2, you get the following:

$$-2x + 6y - 4$$

Notice that the negative sign $(-)$ affects each term inside the parentheses.

PRACTICE

Simplify each algebraic expression. You can check your answers at the end of the section.

16. $4x + 8y + 6x + 3y$

17. $-x(2x - 5) + 7x$

18. $2x(3x - y) + 4x$

19. $12 + 2(6x - 2y) - (2 - 3x)$

20. $4(4x - yx) + 5(x + yx) + 2x(y - 5yx)$

Solving Algebraic Equations

LESSON SUMMARY

In this lesson, you will learn what an algebraic equation is. You will also learn how to solve one-step and multiple-step algebraic equations.

An *algebraic equation* is a math sentence. It always has an equals (=) sign. Algebraic equations say one quantity is equal to another quantity. Here are a few examples of algebraic equations.

$5 + x = 25$

$x - 5 = 25$

$5x = 25$

HOW TO SOLVE ALGEBRAIC EQUATIONS

Solving an algebraic equation means that you have to find the value of the variable. To solve for the value of the variable, you first need to get it alone on one side of the equals sign. This is sometimes called *isolating the variable.*

You know from Lesson 1 that addition and subtraction are *inverse operations*. (They "undo" each other; for example, if 3 + 6 = 9, then 9 − 6 = 3 and 9 − 3 = 6.) Multiplication and division are also inverse operations. (They also "undo" each other; for example, if 30 ÷ 6 = 5, then 5 × 6 = 30.) So, when addition is on the variable side of the equation, you should subtract to isolate the variable. When you see subtraction, you should add. When you see multiplication, you should divide, and when you see division, you should multiply. And if a variable is squared, you should take the square root of it. Got it?

You want to get the variable alone on one side of the equals sign, so you perform mathematical operations to both sides of the equation to isolate the variable. With every step you take to solve the equation, you should ask yourself, "What operations can I use to get the x alone on one side of the equals sign?" But remember, whatever you do to one side of the equation, you have to do to the other. Let's try some examples to see how it works.

Example: Solve the following algebraic expression: $x + 5 = 10$.

Step 1: Ask yourself: What operation is used in the equation? Addition. In order to get the x alone, you will need to get rid of the 5. So, you should use the *inverse operation* and subtract. Remember, you have to perform the operation to *both* sides of the equation.

$x + 5 = 10$

$\underline{\;-5 \;\; -5\;}$ ⟶ Subtract 5 from both sides of the equation.

Step 2: Combine like terms on both sides of the equals sign.

$x + 5 = 10$

$\underline{\;-5 \;\; -5\;}$

$x + 0 = \;5$ ⟶ All like terms have been combined.

Therefore, $x = 5$.

Step 3: You can check your answer by going back to the original equation and plugging in your answer for x. If your answer makes the algebraic expression true, then it's correct. Let's try it out:

$x + 5 = 10$, and $x = 5$, so $5 + 5 = 10$. You know that $10 = 10$, so your answer, $x = 5$ is correct.

Example: Solve the following algebraic expression: $z - 3 = 8$.

Step 1: Ask yourself: What operation is used in the equation? Subtraction. What is the inverse operation of subtraction? Addition. So, add 3 to each side of the equation in order to isolate the variable.

$$\begin{array}{rr} z - 3 = & 8 \\ +3 & +3 \\ \hline \end{array}$$ ⟶ Add 3 to each side of the equation.

Step 2: Combine like terms on both sides of the equals sign.

$$\begin{array}{rr} z - 3 = & 8 \\ +3 & +3 \\ \hline z + 0 = & 11 \\ z = & 11 \end{array}$$ ⟶ All like terms have been combined.

Therefore, $z = 11$.

Check your answer by going back to the original equation and plugging in your answer. $z - 3 = 8$, and z = 11, so 11 – 3 = 8. You know that 8 = 8, so your answer, $z = 11$ is correct.

Example: Solve the following algebraic expression: $5r = -25$.

Step 1: Ask yourself: What operation is used in the equation? Multiplication. What is the inverse operation of this multiplication? Division. So, divide each side of the equation by 5 in order to isolate the variable:

$\frac{5r}{5} = \frac{-25}{5}$ ⟶ Divide each side of the equation by 5

Step 2: Combine like terms on both sides of the equals sign

$\frac{5r}{5} = r$

and

$\frac{-25}{5} = -5$

Therefore, $r = -5$.

Example: Solve the following algebraic expression: $\frac{x}{2} = 12$.

Step 1: Ask yourself: What operation is used in the equation? Division. What is the inverse operation of this division? Multiplication. So, multiply each side of the equation by 2 in order to isolate the variable.

$2 \times \frac{x}{2} = 12 \times 2$ ⟶ Multiply each side of the equation by 2 to isolate the variable.

Step 2: Combine like terms on both sides of the equals sign

$\frac{2x}{2} = x$

and

$12 \times 2 = 24$

Therefore, $x = 24$

PRACTICE

Solve each algebraic equation. You can check your answers that the end of the lesson.

1. $x + 10 = 14$

2. $a - 7 = 12$

3. $y + (-2) = 15$

4. $-r + 3 = 21$

5. $s - 9 = 3$

6. $2x = 12$

7. $-3t = -21$

8. $5q = -45$

9. $\frac{x}{4} = 3$

10. $\frac{x}{3} = 2$

▶ SOLVING MULTIPLE STEP ALGEBRAIC EQUATIONS

The algebraic equations you've solved so far in this lesson have almost all required one inverse operation. But some algebraic equations require more than one inverse operation to isolate the variable and then solve the equation.

When solving multiple-step algebraic equations, you should first add or subtract. Then, multiply or divide. In other words, you follow the order of operations in inverse order! Another way to look at it is to solve for the number attached to the variable last. Let's look at some examples.

Example: Solve the following algebraic expression: $2m + 5 = 13$.

Step 1: First, look for numbers that are being added or subtracted to the term with the variable. In this equation, 5 is added to the $2m$. To simplify, subtract 5 from both sides of the equation.

$$\begin{array}{rcl} 2m + 5 & = & 13 \\ -5 & & -5 \\ \hline 2m & = & 8 \end{array}$$

⟶ Subtract 5 from either side of the equation.

Step 2: Perform the inverse operation for any multiplication or division. The variable in this equation is multiplied by 2, so you must divide each side of the equation by 2 in order to isolate the variable.

$\frac{2m}{2} = \frac{8}{2}$ ⟶ Divide both sides of the equation by 2 to isolate the variable.

Therefore, the answer is $m = 4$.

Many multiple-step algebraic equations can be solved in more than one way. When you check your answers in the answer key, don't worry if you followed slightly different steps to get to the correct answer. The main thing to keep in mind is to add or subtract first. Then, multiply or divide.

Example: Solve the following algebraic expression: $5p + 24 = 3p - 4$.

Step 1: This equation has a variable on each side of the equals sign. So, you first need to get the variable terms on one side of the equation. You can do this by subtracting the smaller of the two variables from each side because they have like terms.

$$5p + 24 = 3p - 4$$

$$\begin{array}{rcl} 5p + 24 & = & 3p - 4 \\ -\,3p \quad\;\; & & -\,3p \\ \hline 2p + 24 & = & -4 \end{array}$$

→ Subtract $3p$ from both sides of the equation.

Step 2: Look for numbers that are being added or subtracted to the term with the variable. In this equation, 24 is added to the $2p$. So you subtract 24 from both sides of the equation.

$$\begin{array}{rcl} 2p + 24 & = & -4 \\ -\,24 & & -\,24 \\ \hline 2p \quad\;\; & = & -28 \end{array}$$

→ Subtract 24 from both sides of the equation to isolate the variable.

Step 3: Perform the inverse operation for any multiplication or division. The variable in this equation is multiplied by 2, so you divide each side of the equation by 2 to isolate the variable.

$\frac{2p}{2} = \frac{-28}{2}$ → Divide both sides of the equation by 2 to isolate the variable.

Therefore, the answer is $p = -14$.

Example: Solve the following algebraic expression: $4(b + 1) = 20$.

Step 1: This equation has parentheses, so you need to remove the parentheses by distributing the 4.

$4(b + 1) = 4b + 4$

So, your equation becomes

$4b + 4 = 20$

Step 2: Look for numbers that are being added or subtracted to the term with the variable. Subtract 4 from both sides of the equation.

$$\begin{array}{rcl} 4b + 4 & = & 20 \\ -\,4 & & -\,4 \\ \hline 4b \quad\;\; & = & 16 \end{array}$$

→ Subtract 4 from both sides of the equation.

Step 3: Perform the inverse operation for any multiplication or division. Divide each side of the equation by 4.

$\frac{4b}{4} = \frac{16}{4}$ ⟶ Divide both sides of the equation by 4.

Therefore, your answer is $b = 4$.

PRACTICE

Solve each algebraic equation. You can check your answers at the end of the lesson.

11. $3a + 4 = 13$

12. $2p + 2 = 16$

13. $4(c - 1) = 12$

14. $3x - 4 = 2x + 4$

15. $10w + 14 - 8w = 12$

16. $5(b + 1) = 60$

17. $3(y - 9) - 2 = -35$

18. $4q + 12 = 16$

19. $\frac{t}{5} - 3 = -9$

20. $6(2 + f) = 5f + 15$

Real World Problems

These problems apply the skills you've learned in Section 7 to everyday situations. As you work through these problems, you'll see that the skills you've learned in this section aren't only important for math tests. They are important skills for questions that come up every day that you need to solve. Don't forget to use all the skills you've learned in previous chapters translating word sentences to math sentences. Also, refer to the table *Translating Math Relationships into Algebraic Expressions* on p. 245 if you need to refresh your memory.

Algebraic expressions and equations are very useful in solving the types of word problems you are likely to find on math tests. Here are some tips for solving these types of problems.

- **First, read the problem to determine what you are looking for.**
- **Then, write the amount you are looking for in terms of *x* (or whatever letter you want to use). You can do this by writing "Let *x* = ..." Write any other unknown amounts in terms of *x*, too.**
- **Lastly, set up the algebraic expressions in an equation with an equals sign and solve for the variable.**

1. An adult ticket to the cinema costs twice as much as a child's ticket. Timothy paid $28 for two adult tickets and three children's tickets. Write an algebraic expression you could use to find the price of an adult's ticket to the cinema.

2. Jeremy and Mai wanted to see who could do the most sit-ups in one minute. Mai did 12 more sit-ups than Jeremy did. The total number of sit-ups both Jeremy and Mai did was 66. How many sit-ups did Jeremy do?

3. There are 12 fewer girls than twice the number of boys signed up for Rolling Hills Tennis Camp. If 60 girls and boys are signed up for the camp, how many boys are signed up?

4. The ratio of blue jellybeans to red jellybeans to yellow jellybeans is 3:2:1. If there are 12,000 jellybeans total, how many are red jellybeans?

5. Ten times 40% of a number is equal to 4 less than the product of 6 times the number.

6. Peter got a raise on his after school job. He now makes $10 per hour, which is $4 more than $\frac{2}{3}$ his original hourly rate. How much did Peter make before he got his raise? How much was his raise?

SECTION

7

Answers & Explanations

LESSON 16

1. $10 - 2x$
2. $6 - x$
3. $(5 + x) - 11$
4. $7x + 3$
5. $4x + 2$
6. $\frac{x}{2}$ or $x \div 2$
7. $8x - 1$
8. $10 - (14 + x)$
9. $3xy$
10. $\frac{5}{xy}$ or $5 \div xy$
11. **a.** Begin by replacing the variable with the number 1: $1 + 2(1) + 10$. Then, solve.
$1 + 2(1) + 10 = 13$
The answer is 13.
b. Begin by replacing the variable with -1: $-1 + 2(-1) + 10$. Then, solve.
$-1 + 2(-1) + 10 =$
$-3 + 10 = 7$
The answer is 7.
c. Begin by replacing the variable with 5: $5 + 2(5) + 10$. Then, solve.
$5 + 2(5) + 10 =$
$5 + 10 + 10 = 25$
The answer is 25.

12. a. Begin by replacing the variable with 1: $(1)^2 + 10$. Then, solve.

$(1)^2 + 10 =$

$1 + 10 = 11$

The answer is 11.

b. Begin by replacing the variable with –1: $(-1)^2 + 10$. Then, solve.

$(-1)^2 + 10 =$

$1 + 10 = 11$

The answer is 11.

c. Begin by replacing the variable with 3: $(3)^2 + 10$. Then, solve.

$(3)^2 + 10 =$

$9 + 10 = 19$

The answer is 19.

13. a. Begin by replacing the variable with –1: $2\,[(-1) - 1]$. Then, solve.

$2[(-1) - 1] =$

$2(-2) = -4$

The answer is –4.

b. Begin by replacing the variable with –2: $2\,[(-2) - 1]$. Then, solve.

$2[(-2) -1] =$

$2(-3) = -6$

The answer is –6.

c. Begin by replacing the variable with +3: $2\,(3 - 1)$. Then, solve.

$2(3 - 1) =$

$2(2) = 4$

The answer is 4.

14. a. Begin by replacing the x with –1 and the y with +1: $5[(-1) - 3] + 2(1)$. Then, solve.

$5[(-1) - 3] + 2(1) =$

$5(-4) + 2 =$

$-20 + 2 = -18$

The answer is –18.

b. Begin by replacing the x with 3 and the y with –2: $5(3 - 3) + 2(-2)$. Then, solve.

$5(3 - 3) + 2(-2) =$

$5(0) + (-4) =$

$0 + (-4) = -4$

The answer is –4.

c. Begin by replacing the x with 5 and the y with –1: $5(5 - 3) + 2(-1)$. Then, solve.

$5(5 - 3) + 2(-1) =$

$5(2) + (-2) =$

$10 + (-2) = 8$

The answer is 8.

15. a. Begin by replacing the x with -1 and the y with 1: $3(-1) + \frac{2(1)}{3}$. Then, solve.

$3(-1) + \frac{2(1)}{3} =$

$-3 + \frac{2}{3} = -2\frac{1}{3}$

The answer is $-2\frac{1}{3}$.

b. Begin by replacing the x with 5 and the y with -3: $3(5) + \frac{2(-3)}{3}$. Then, solve.

$3(5) + \frac{2(-3)}{3} =$

$15 + \frac{-6}{3} =$

$15 + -2 = 13$

The answer is 13.

c. Begin by replacing the x with 12 and the y with -12: $3(12) + \frac{2(-12)}{3}$. Then, solve.

$3(12) + \frac{2(-12)}{3} =$

$36 + \frac{-24}{3} =$

$36 + -8 =$

The answer is 28.

16. Begin by writing the like terms next to each other.

$4x + 6x + 8y + 3y$

Then, add the like terms together.

$10x + 11y$

The answer is $10x + 11y$.

17. First, distribute to eliminate the parentheses.

$-x(2x - 5) + 7x = -2x^2 + 5x + 7x$

Then, combine the like terms.

$-2x^2 + 5x + 7x = -2x^2 + 12x$

The answer is $-2x^2 + 12x$.

18. First, distribute to eliminate the parentheses.

$2x(3x - y) + 4x = 6x^2 - 2xy + 4x$

There are no like terms to combine, so the answer is $6x^2 - 2xy + 4x$.

19. First, distribute to eliminate the parentheses.

$12 + 2(6x - 2y) - (2 - 3x)$

$12 + 12x - 4y - 2 + 3x$

Then, write the like terms next to each other.

$12 - 2 + 12x + 3x - 4y$

Finally, combine the like terms.

$10 + 15x - 4y$

The answer is $10 + 15x - 4y$.

20. First, distribute to eliminate the parentheses.
$4(4x - yx) + 5(x + yx) + 2x(y - 5yx) = 16x - 4yx + 5x + 5yx + 2xy - 10yx^2$

Then, write the like terms next to each other. Note that because multiplication has the commutative property (two or more factors can be multiplied in any order without changing the product), yx and xy are the same term, even though they are in different order.
$16x + 5x - 4yx + 5yx + 2xy - 10yx^2$

Finally, combine the like terms.

$21x + 3yx - 10yx^2$

The answer is $21x + 3yx - 10yx^2$.

LESSON 17

1. Begin by asking yourself what operation is used in the equation: addition. Then, perform the inverse operation (subtraction) to both sides of the equation:
$x + 10 = 14$

$$\begin{array}{r} x + 10 = \ \ 14 \\ -10 \ \ \ -10 \\ \hline \end{array}$$

Finally, combine like terms and solve for the variable.

$x + 0 = 4$
$x = 4$

Thus, the final answer is $x = 4$.

2. Begin by asking yourself what operation is used in the equation: subtraction. Then, perform the inverse operation (addition) to both sides of the equation:
$a - 7 = 12$

$$\begin{array}{r} a - 7 = \ \ 12 \\ +7 \ \ \ +7 \\ \hline \end{array}$$

Finally, combine like terms and solve for the variable.

$a + 0 = 19$
$a = 19$

The final answer is $a = 19$.

3. Begin by asking yourself what operation is used in the equation: addition. Then, perform the inverse operation (subtraction) to both sides of the equation:

$y + (-2) = 15$

$$\begin{array}{r} y + (-2) = 15 \\ \underline{\quad +2 \quad +2} \end{array}$$

Finally, combine like terms and solve for the variable.

$y + 0 = 17$
$y = 17$

Thus, the final answer is $y = 17$.

4. Begin by asking yourself what operation is used in the equation: addition. Then, perform the inverse operation (subtraction) to both sides of the equation:

$-r + 3 = 21$

$$\begin{array}{r} -r + 3 = 21 \\ \underline{\quad -3 \quad -3} \end{array}$$

Finally, combine like terms and solve for the variable.

$-r + 0 = 18$
$-r = 18$

But you want to solve for r; not $-r$. So, multiply each side by -1.

$-r \times -1 = 18 \times -1$
$r = -18$

The final answer is $r = -18$.

5. Begin by asking yourself what operation is used in the equation: subtraction. Then, perform the inverse operation (addition) to both sides of the equation:

$s - 9 = 3$

$$\begin{array}{r} s - 9 = \ 3 \\ \underline{+9 \quad +9} \end{array}$$

Finally, combine like terms and solve for the variable.

$s + 0 = 12$
$s = 12$

The final answer is $s = 12$.

6. Begin by asking yourself what operation is used in the equation: multiplication. Then, perform the inverse operation (division) to both sides of the equation:

$2x = 12$

$\frac{2x}{2} = \frac{12}{2}$

Finally, combine like terms and solve for the variable.

$x = 6$

The final answer is $x = 6$.

7. Begin by asking yourself what operation is used in the equation: multiplication. Then, perform the inverse operation (division) to both sides of the equation:

$-3t = -21$

$\frac{-3t}{3} = \frac{-21}{-3}$

Finally, combine like terms and solve for the variable.

$t = 7$

The final answer is $t = 7$.

8. Begin by asking yourself what operation is used in the equation: multiplication. Then, perform the inverse operation (division) to both sides of the equation:

$5q = -45$

$\frac{5q}{5} = \frac{-45}{5}$

Finally, combine like terms and solve for the variable.

$q = -9$

The final answer is $q = -9$.

9. Begin by asking yourself what operation is used in the equation: division. Then, perform the inverse operation (multiplication) to both sides of the equation:

$\frac{x}{4} = 3$

$4 \times \frac{x}{4} = 3 \times 4$

Finally, combine like terms and solve for the variable.

$x = 12$

The final answer is $x = 12$.

10. Begin by asking yourself what operation is used in the equation: division. Then, perform the inverse operation (multiplication) to both sides of the equation:

$$\frac{x}{3} = 2$$
$$3 \times \frac{x}{3} = 2 \times 3$$

Finally, combine like terms and solve for the variable.

$$x = 6$$

The final answer is $x = 6$.

11. Begin by performing the inverse operation for addition:
$3a + 4 = 13$

$$\begin{array}{rr} 3a + 4 = & 13 \\ -4 & -4 \\ \hline 3a + 0 = & 9 \end{array}$$

$$3a = 9$$

Then, perform the inverse operation for multiplication and solve for a.

$$\frac{3a}{3} = \frac{9}{3}$$
$$a = 3$$

The final answer is $a = 3$.

12. Begin by performing the inverse operation for addition.
$2p + 2 = 16$

$$\begin{array}{rr} 2p + 2 = & 16 \\ -2 & -2 \\ \hline 2p \quad = & 14 \end{array}$$

Then, perform the inverse operation for multiplication and solve for p.

$$\frac{2p}{2} = \frac{14}{2}$$
$$p = 7$$

The final answer is $a = 7$.

13. First, eliminate the parentheses by multiplying (distributing the 4).

$4(c - 1) = 12$

$4c - 4 = 12$

Then, perform the inverse operation for subtraction.

$$\begin{array}{rcr} 4c - 4 & = & 12 \\ +4 & & +4 \\ \hline 4c & = & 16 \end{array}$$

Finally, perform the inverse operation for multiplication and solve for c.

$\frac{4c}{4} = \frac{16}{4}$

$c = 4$

The final answer is $c = 4$.

14. First, group the variables on one side of the equation by subtracting the smaller of the two variables from both sides.

$3x - 4 = 2x + 4$

$$\begin{array}{rcr} 3x - 4 & = & 2x + 4 \\ -2x & & -2x \\ \hline x - 4 & = & 4 \end{array}$$

Then, perform the inverse operation for subtraction and solve for x.

$$\begin{array}{rcr} x - 4 & = & 4 \\ +4 & & +4 \\ \hline x & = & 8 \end{array}$$

Thus, the final answer is $x = 8$.

15. Begin by grouping like terms together.

$10w + 14 - 8w = 12$

$10w - 8w + 14 = 12$

Combine the like terms.

$10w - 8w = 2w$

$2w + 14 = 12$

Then, perform the inverse operation for addition.

$$\begin{array}{rcr} 2w + 14 & = & 12 \\ -14 & & -14 \\ \hline 2w + 0 & = & -2 \end{array}$$

$2w = -2$

Finally, perform the inverse operation for multiplication and solve for w.

$\frac{2w}{2} = \frac{-2}{2}$

$w = -1$

The final answer is $w = -1$.

16. First, eliminate the parentheses by multiplying (distributing the 5).

$5(b + 1) = 60$

$5b + 5 = 60$

Then, perform the inverse operation for addition.

$$\begin{array}{rr} 5b + 5 = & 60 \\ -5 & -5 \\ \hline 5b + 0 = & 55 \end{array}$$

$5b = 55$

Finally, perform the inverse operation for multiplication and solve for b.

$\frac{5b}{5} = \frac{55}{5}$

$b = 11$

The final answer is $b = 11$.

17. First, eliminate the parentheses by multiplying (distribute the 3).

$3(y - 9) - 2 = -35$

$3y - 27 - 2 = -35$

Combine the like terms.

$3y - 29 = -35$

Then, perform the inverse operation for subtraction.

$$\begin{array}{rr} 3y - 29 = & -35 \\ +29 & +29 \\ \hline 3y \quad = & -6 \end{array}$$

Finally, perform the inverse operation for multiplication and solve for y.

$\frac{3y}{3} = \frac{-6}{3}$

$y = -2$

The final answer is $y = -2$.

18. Begin by performing the inverse operation for addition.

$$\begin{array}{rcr} 4q + 12 &=& 16 \\ -12 && -12 \\ \hline 4q &=& 4 \end{array}$$

Then, perform the inverse operation for multiplication and solve for q.

$\frac{4q}{4} = \frac{4}{4}$

$q = 1$

The final answer is $q = 1$.

19. Begin by performing the inverse operation for subtraction.

$\frac{t}{5} - 3 = -9$

$$\begin{array}{rcr} \frac{t}{5} - 3 &=& -9 \\ +3 && +3 \\ \hline \frac{t}{5} &=& -6 \end{array}$$

Then, perform the inverse operation for division and solve for t.

$5 \times \frac{t}{5} = -6 \times 5$

$t = -30$

The final answer is $t = -30$.

20. First, eliminate the parentheses by multiplying (distribute the 6).

$6(2 + f) = 5f + 15$

$12 + 6f = 5f + 15$

Then, group the variables on one side of the equation.

$$\begin{array}{rcr} 12 + 6f &=& 5f + 15 \\ -5f && -5f \\ \hline 12 + f &=& 15 \end{array}$$

Finally, perform the inverse operation for addition and solve for f.

$$\begin{array}{rcr} 12 + f &=& 15 \\ -12 && = -12 \\ \hline f &=& 3 \end{array}$$

The final answer is $f = 3$.

REAL WORLD PROBLEMS

1. Your task is to translate the words into an algebraic equation. First, because you are looking for the price of an *adult* ticket, let x = the price of an adult's ticket to the cinema. Therefore, a child's ticket costs $\frac{1}{2}x$. If Timothy paid \$28 for two adult tickets and three children's tickets, then $2x + 3(\frac{1}{2}x) = \$28$. To calculate the cost of an adult's ticket, you would solve for x; however, the question only asks you to set up the equation, so your final answer is $3x + 2(2x) = \$28$.

2. First, translate the words into an algebraic equation. Let x = the number of sit-ups Jeremy did. Then, you know that Mai did $x + 12$ sit-ups. If you add these two amounts together, you will get the total number of sit-ups done by both people. So, $x + x + 12 = 66$. Simplify: $2x + 12 = 66$.

To solve, first subtract 12 from each side of the equation.

$$\begin{array}{rcr} 2x + 12 & = & 66 \\ -\,12 & & -\,12 \\ \hline 2x \quad\;\; & = & 54 \end{array}$$

Then, divide each side by 2.

$2x \div 2 = 54 \div 2$
$x = 27$

So, Jeremy did 27 sit-ups.

3. First, translate the words into an algebraic equation. Let x = the number of boys signed up for the camp. Then, you know that the number of girls signed up is $2x - 12$. If you add these two amounts together, you will get the total number of people signed up for the camp (60). So, $x + 2x - 12 = 60$. Simplify: $3x - 12 = 60$. To solve, first add 12 to each side of the equation.

$$\begin{array}{rcr} 3x - 12 & = & 60 \\ +\,12 & & +\,12 \\ \hline 3x \quad\;\; & = & 72 \end{array}$$

Then, divide each side by 3.

$3x \div 3 = 72 \div 3$
$x = 24$

So, 24 boys are signed up for Rolling Hills Tennis Camp.

4. We can apply the 3:2:1 ratio to an algebraic formula: $3x + 2x + 1x$ = total number of jellybeans. The $3x$ represents the blue jellybeans, the $2x$ represents the red jellybeans, and the $1x$ represents the green jellybeans. We know that the total number of jellybeans = 12,000, so we have:

$3x + 2x + 1x = 12{,}000$

$6x = 12{,}000$

$x = 2{,}000$

Remember, we want "red jellybeans" = $2x = 2 \times 2{,}000 = 4{,}000$.

5. Let's call our number "N." Looking at the statement, the first part says "Ten times 40% of a number . . . ". We can express this mathematically as $10 \times .40N$. Remember that 40% is the same as .40 and "of" means *multiply*. The next part of the statement is " . . . is equal to 4 less than the product of 6 times the number." This can be expressed as $= 6N - 4$. Putting it all together, we have $10 \times .40N = 6N - 4$, or $4N = 6N - 4$. We add 4 to both sides to get $4 + 4N = 6N$. Next, subtract $4N$ from both sides to yield $4 = 2N$, thus, $N = 2$.

6. First, translate the words into an algebraic equation. Let x = Peter's original hourly rate. Set up the equation:

$4 + \frac{2}{3}x = \$10$

Now, solve the equation as you would any other algebraic equation.

Subtract 4 from both sides:

$$\begin{array}{rcr} 4 + \frac{2}{3}x & = & 10 \\ -4 & & -4 \\ \hline \frac{2}{3}x & = & 6 \end{array}$$

Multiply both sides by $\frac{3}{2}$:

$\frac{3}{2} \times \frac{2}{3}x = 6 \times \frac{3}{2}$

$x = 9$

So, Peter's original hourly rate was \$9. His raise was \$10 – \$9 = \$1 per hour.

SECTION

8

Introduction to Geometry

THIS SECTION WILL introduce you to the basics of geometry. Everywhere you look, you can see geometric shapes—buildings, cars, and other objects around you are combinations of shapes such as squares, rectangles, triangles, and circles. You'll learn important geometry vocabulary in this section so that you'll understand the language of geometry. You'll also learn important facts to help you see how geometry fits into your everyday life.

Points, Lines, Rays, Line Segments, and Angles

LESSON SUMMARY

This lesson introduces the basic vocabulary of geometry. You will learn about points, lines, rays, line segments, and angles.

Take a moment to look around you, and you'll see geometry everywhere. The pages of this book, your computer screen, and your desk are all rectangular. Keep looking around and you'll find triangles, circles, and other geometric shapes. In this lesson, you'll learn some basic words in the language of geometry.

BREAKING THINGS DOWN

You are probably already familiar with many two-dimensional and three-dimensional shapes, but geometry involves more than squares and cubes. Let's begin with the basics—the pieces that make these shapes up.

SOME BASIC GEOMETRY VOCABULARY

NAME	FIGURE	SYMBOL	READ AS	PROPERTIES	EXAMPLES
Point	• A	• A	point A	■ has no size ■ has no dimension ■ indicates a definite location ■ named with an uppercase letter	■ pencil point ■ corner of a room
Line	A B ⟷ ℓ ⟷	$\overleftrightarrow{AB}$ or $\overleftrightarrow{BA}$	line AB or BA or line ℓ	■ is straight ■ has no thickness ■ an infinite set of points that extends in opposite directions ■ one dimension	■ highway without boundaries ■ hallway without bounds
Ray	A B ⟶	$\overrightarrow{AB}$	ray AB (endpoint is always first)	■ is part of a line ■ has only one endpoint ■ an infinite set of points that extends in one direction ■ one dimension	■ flashlight ■ laser beam
Line segment	A B —	$\overline{AB}$ or $\overline{BA}$	segment AB or BA	■ is part of a line ■ has two endpoints ■ one dimension	■ edge of a ruler ■ base board

Example: Write a symbol to represent the following figure.

After examining the table, you know that this is a line because it has an arrow on each end. There are two ways to represent this line in math symbols:

$\overleftrightarrow{ST}$

$\overleftrightarrow{TS}$

Example: How many different line segments are shown in the figure below? (If you don't remember what a line segment is, refer back to the table on page 278.) Write a symbol to represent each line segment.

There are three line segments here. They are

$\overline{DE}$ $\overline{EF}$ $\overline{DF}$

PRACTICE

Write a symbol to represent each figure.

1.

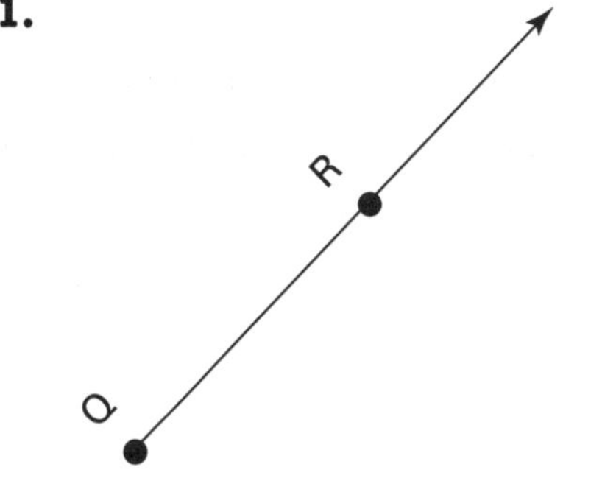

2. M N

3. G H

4. W Z

5. R Q

Use the information in the table on page 278 to answer the following questions.

6. Write in words how you would read this figure: • P

7. Look at the figure below. Then, answer the questions.

a. List as many different ways as you can think of to represent this line with symbols. (Hint: You should find six.)

b. Name two different rays that start with the endpoint Y. Write your answer using symbols.

WHAT ARE ANGLES?

Now that you know about points, lines, rays, and line segments, you are ready to learn about angles. An *angle* is formed by two rays that share a common endpoint. One ray forms each side of the angle and the common endpoint is called the *vertex*. The figure below is an example of an angle.

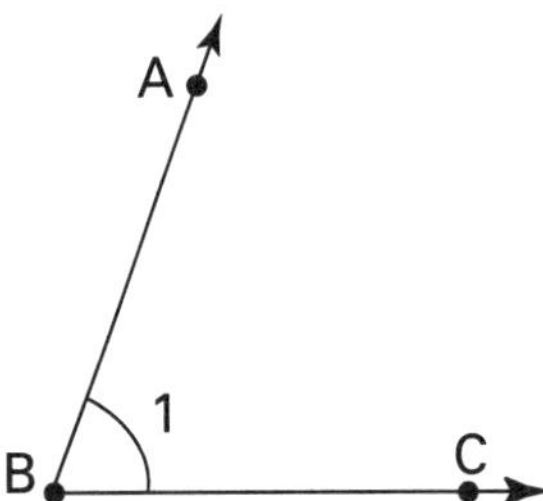

There is more than one way to name an angle. The symbol for an angle is ∠. Here are some ways to name an angle.

- **You can use three letters to name an angle.** The middle letter represents the vertex. For example, the angle above could be named ∠ABC or ∠CBA. In both cases, the B is in the middle because it represents the vertex, or the endpoint, of the two rays that make up the angle.
- **You can use only one letter to name an angle** if the angle you are naming does not share its vertex point with another angle. For example, the angle is alone. It does not share its vertex point with any other angle. So, you could name it ∠B.
- **You can use a number.** Look at the angle above again. Notice that a number is written inside the angle. Sometimes angles will be numbered in this way. When this number does not represent the measurement of the angle (we'll discuss that in just a minute), then you can use the number to name the angle: ∠1.

Let's quickly review how to name an angle.

NAMING ANGLES

RULE	HOW TO USE THE RULE	EXAMPLE
Use the angle symbol with three letters	The middle letter should represent the vertex	∠ABC ∠CBA
Use the angle symbol with the vertex letter	Only if the angle you are naming does not share its vertex point with another angle	∠B
Use the angle symbol with a number	Only if the number isn't the measurement of the angle	∠1

PRACTICE

8. Break each angle into its vertex and rays. Write the vertex and rays in symbols.

a.

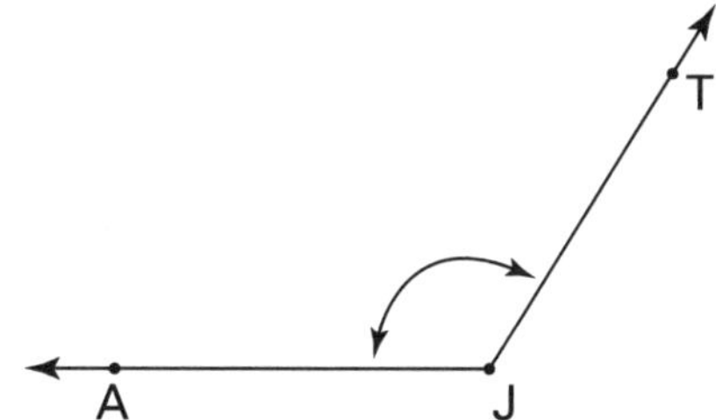

b.

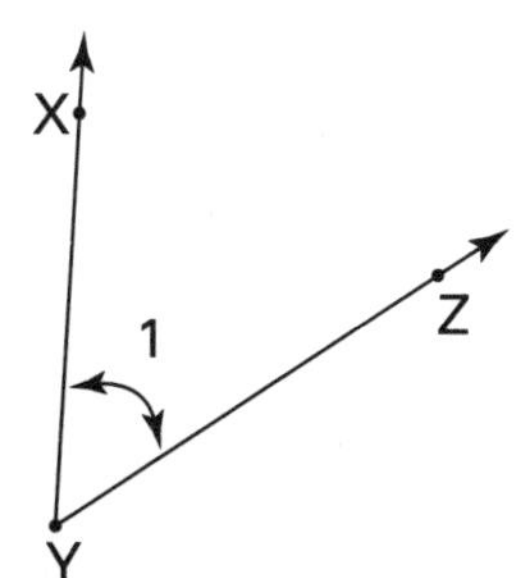

9. Give four different names for the following angle.

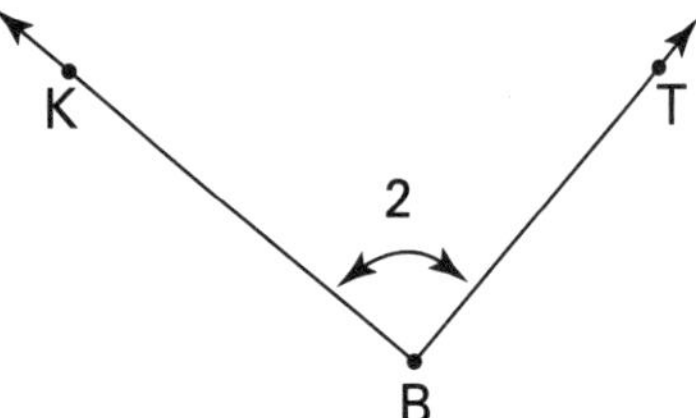

MEASURING ANGLES

Angles are measured in a unit called the *degree*. The symbol for a degree is °. You can use an instrument called a *protractor* to measure an angle. Most protractors look like a clear plastic ruler with a rounded edge like the one shown below.

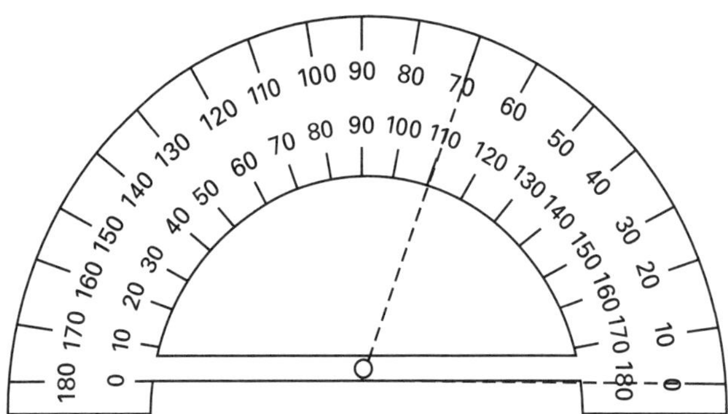

Here's how to use a protractor to measure an angle.

Step 1: Place the cross mark or hole in the middle of the straight edge of the protractor on the vertex of the angle you wish to measure.

Step 2: Line up one of the angle's rays with one of the zero marks on the protractor. (Note that there are two zero marks on the protractor. It doesn't matter which one you use.)

Step 3: Keeping the protractor in place, trace the length of the angle's other ray out to the curved part of the protractor.

Step 4: Read the number closest to the ray. (Notice that there are two rows of numbers on the protractor. Make sure you are measuring from zero.) This is the measure of the angle in degrees.

Example: Measure the following angle using a protractor.

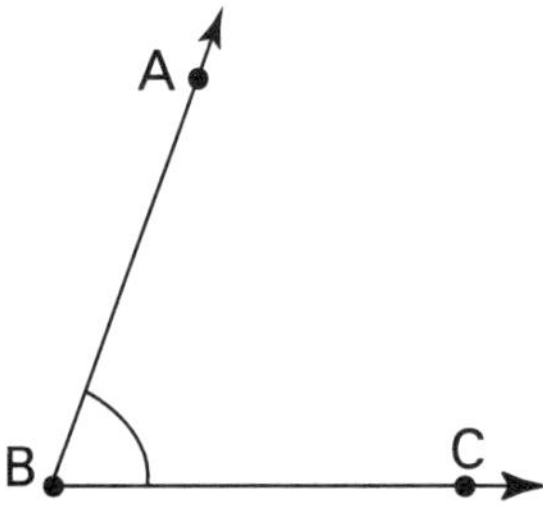

Step 1: Place the cross mark or hole in the middle of the straight edge of the protractor on the vertex B.
Step 2: Line up $\overrightarrow{BC}$ with the zero mark on the right side of the protractor.
Step 3: Keeping the protractor in place, trace the length of $\overrightarrow{BA}$ out to the curved part of the protractor.
Step 4: Read the number closest to $\overrightarrow{BA}$ as shown below.

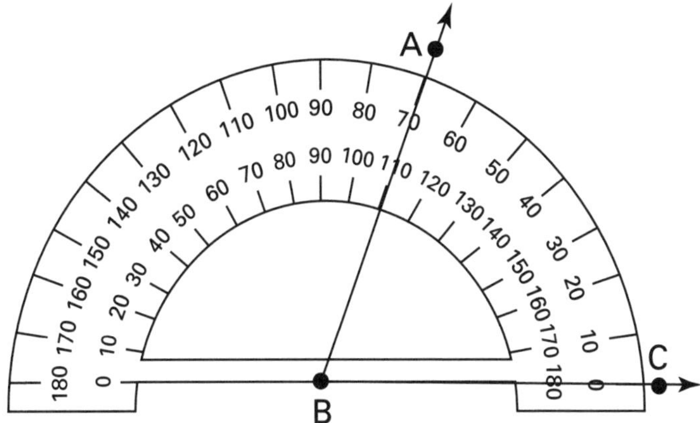

∠ABC is 70°.

When measuring an angle whose rays are shorter than the distance to the numbers on your protractor, use a ruler to carefully extend the lines of the rays. If you keep the rays straight, your measurement will be accurate. The length of the rays does not affect the measure of the angle.

PRACTICE

10. Use a protractor to measure the following angles.

a.

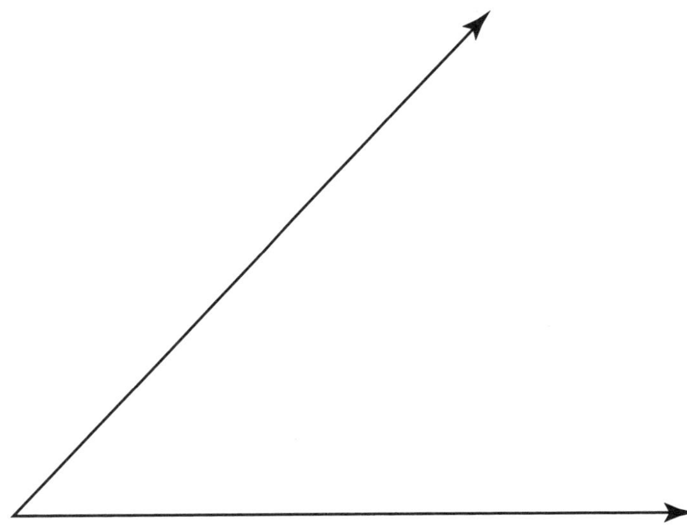

b.

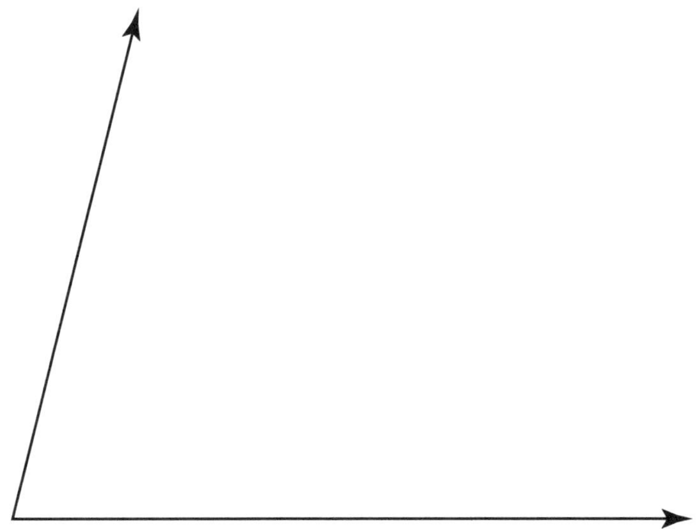

c.

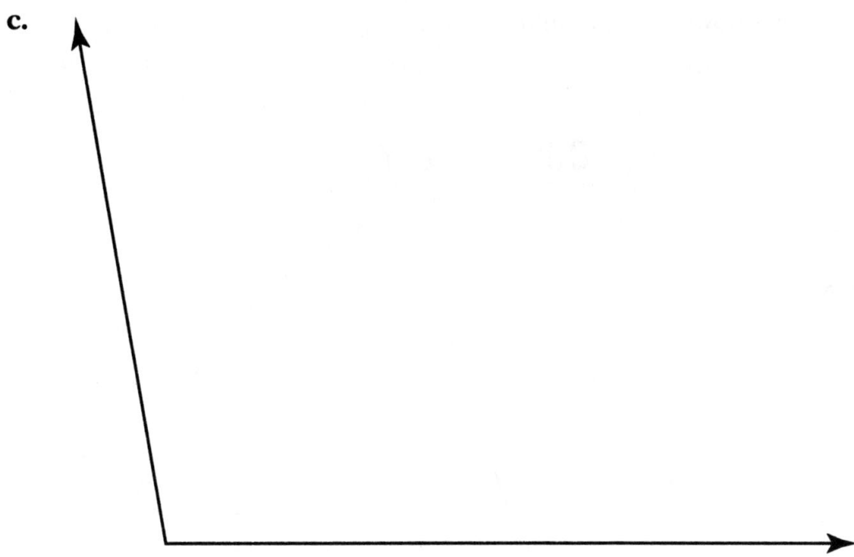

d.

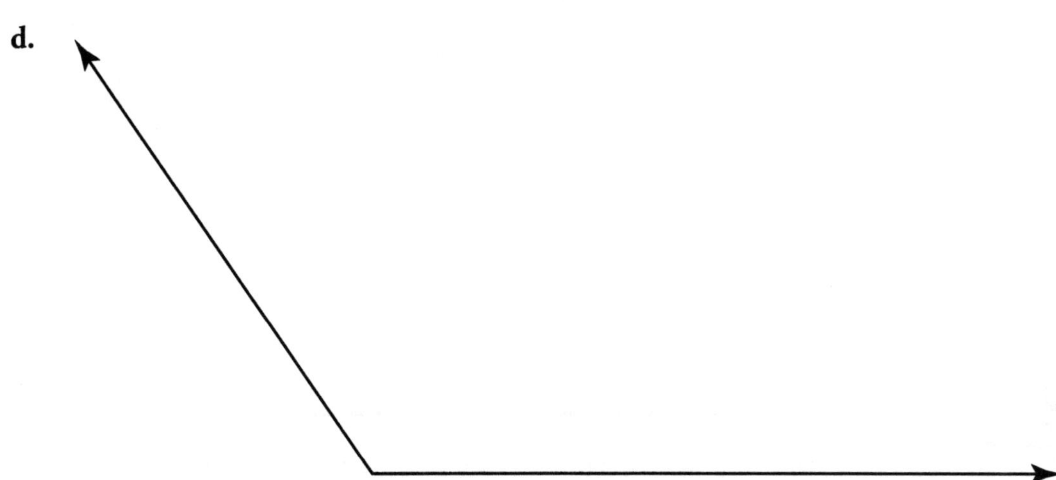

11. Give the measure of the following angles in the figure.

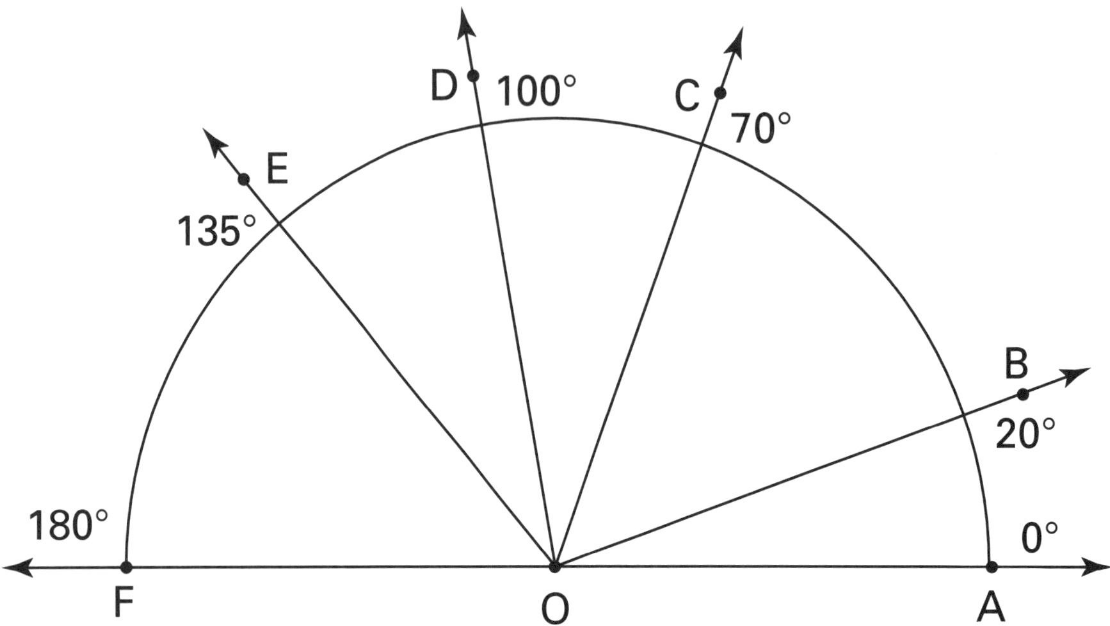

a. ∠AOC
b. ∠AOF
c. ∠BOD
d. ∠COE

You can also use a protractor to draw angles of a certain size. First, draw a ray. Then, place the cross mark or hole in the middle of the straight edge of the protractor on the ray's endpoint. Find the degree of the angle you wish to draw along the curved edge of the protractor. Make sure you are measuring from zero. Mark the point on your paper next to the number of degrees you want your angle to be and, using a ruler, connect the first ray's endpoint to the dot.

▶ CLASSIFYING ANGLES

Angles are classified by their measures. They can be acute, obtuse, right, straight, and reflex. Here are some examples of these different kinds of angles.

Types of Angles

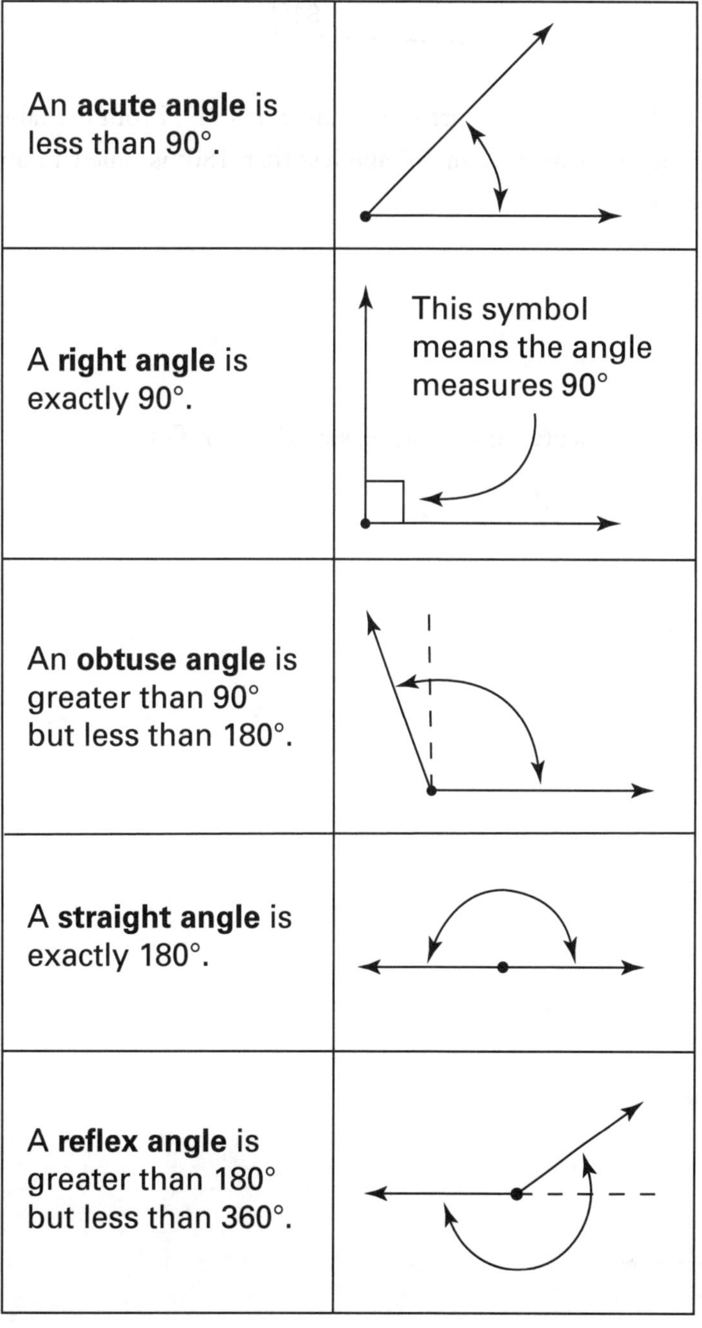

An **acute angle** is less than 90°.	
A **right angle** is exactly 90°.	This symbol means the angle measures 90°
An **obtuse angle** is greater than 90° but less than 180°.	
A **straight angle** is exactly 180°.	
A **reflex angle** is greater than 180° but less than 360°.	

Example: Classify the following angle using the terms in the table on page 287.

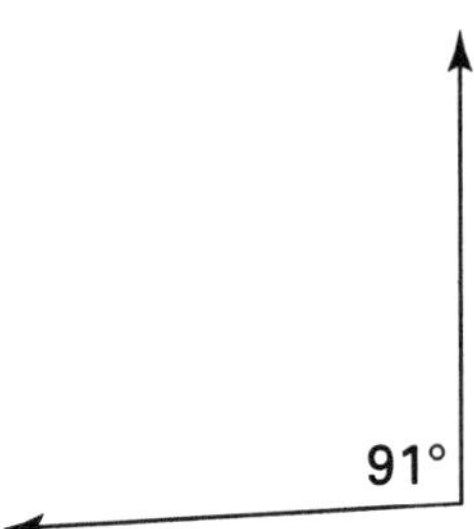

Begin by looking at the measure of the angle. It's 91°. From the table, you know that an angle that measures more than 90° and less than 180° is called an obtuse angle. So the angle is obtuse.

PRACTICE

Classify each angle as either acute, right, obtuse, straight, or reflex.

12.

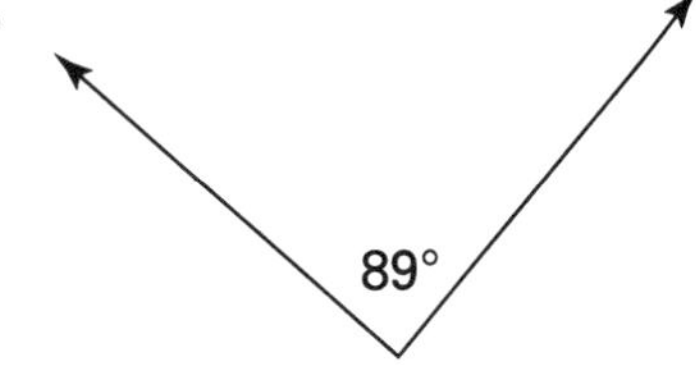

13.

180°

14.

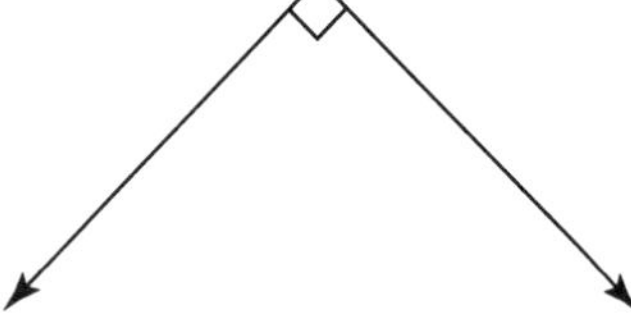

15.

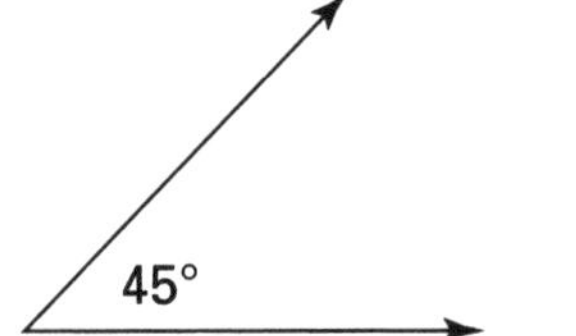

16.

150°

17.

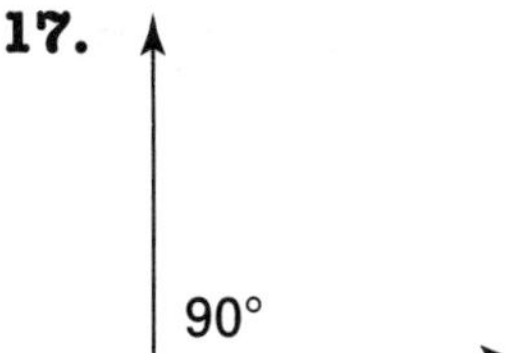

RELATIONSHIPS BETWEEN LINES AND ANGLES

Lines have different relationships with one another depending on whether and how they intersect to form angles. Here are some examples of different relationships lines can have with one another.

LINE RELATIONSHIPS

NAME	DEFINITION	EXAMPLE
Parallel lines	Two lines on a flat surface (also called a plane in geometry) that never intersect The symbol ∥ indicates that two lines are parallel.	p q $p \parallel q$
Perpendicular lines	Two lines that intersect and form right angles The symbol ⊥ indicates that two lines are perpendicular.	k m $k \perp m$
Transversal lines	A line that cuts across two or more lines at different points	t m n transversal

When a transversal line intersects two parallel lines, eight angles form as shown in the figure below.

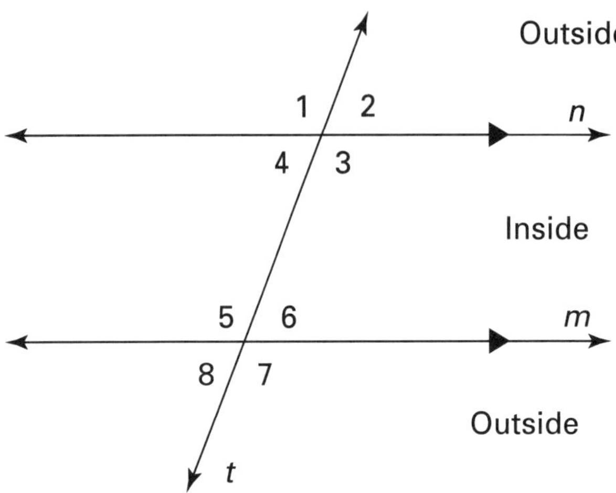

We call these angles different names, depending upon their relationships both to each other and to parallel lines. For example, some pairs of angles formed by a transversal and two parallel lines are said to be *congruent*. Congruent angles are equal in measure. Some pairs of angles formed by a transversal and two parallel lines are said to be *supplementary*. The measures of supplementary angles add up to 180°. Here is a summary of these relationships.

ANGLES FORMED BY PARALLEL LINES AND A TRANSVERSAL

NAME	DEFINITION	RELATIONSHIP	EXAMPLE
Vertical angles	Angles directly across or opposite from each other	Vertical angles are congruent	Angles 1 and 3 Angles 2 and 4 Angles 5 and 7 Angles 6 and 8

NAME	DEFINITION	RELATIONSHIP	EXAMPLE
Corresponding angles	Angles on the same side of the transversal and either both above or both below the parallel lines	Corresponding angles are congruent	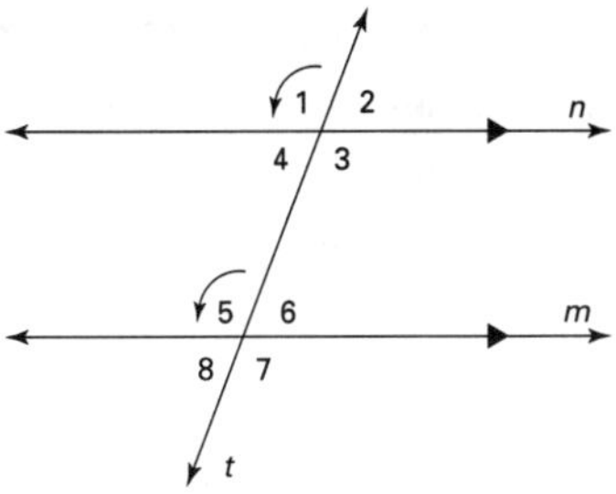 1, 2, n, 4, 3, 5, 6, m, 8, 7, t Angles 1 and 5 Angles 2 and 6 Angles 3 and 7 Angles 4 and 8
Alternate interior angles	Angles inside the parallel lines and on opposite sides of the transversal	Alternate interior angles are congruent	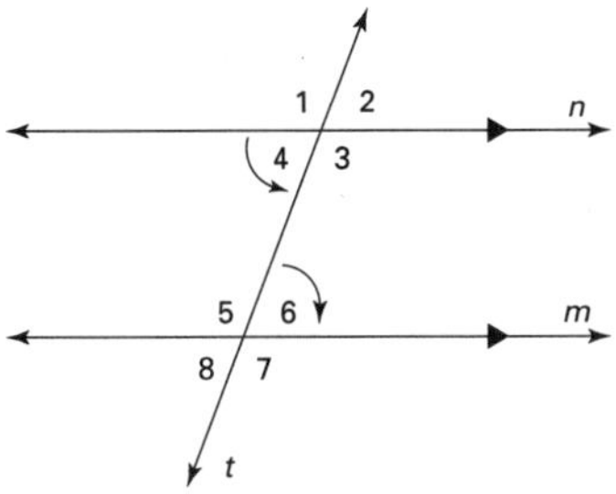 1, 2, n, 4, 3, 5, 6, m, 8, 7, t Angles 4 and 6 Angles 3 and 5
Alternate exterior angles	Angles outside the parallel lines and on opposite sides of the transversal	Alternate exterior angles are congruent	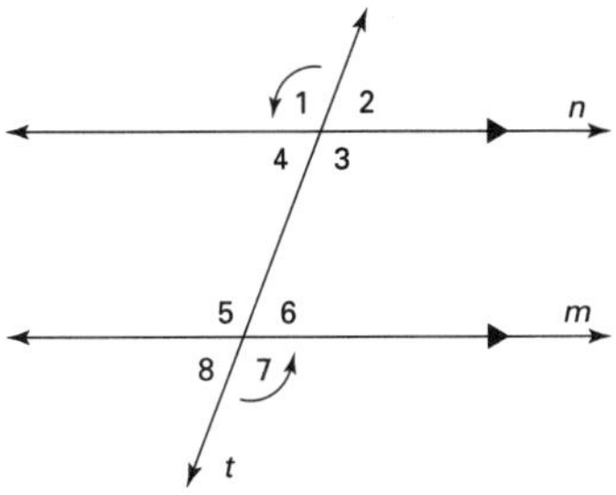 1, 2, n, 4, 3, 5, 6, m, 8, 7, t Angles 1 and 7 Angles 2 and 8
Same side interior angles	Angles inside the parallel lines and on the same side of the transversal	Same side interior angles are supplementary	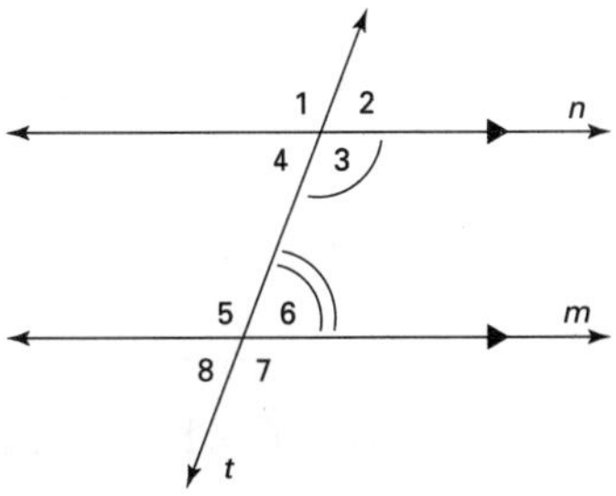 1, 2, n, 4, 3, 5, 6, m, 8, 7, t Angles 3 and 6 Angles 4 and 5

NAME	DEFINITION	RELATIONSHIP	EXAMPLE
Same side exterior angles	Angles outside the parallel lines and on the same side of the transversal	Same side interior angles are supplementary	Angles 2 and 7 Angles 1 and 8

You can use this information to solve many kinds of problems in geometry. Let's look at an example.

Example: Look at the figure below. If the measure of ∠1 is 115°, what is the measure of ∠6?

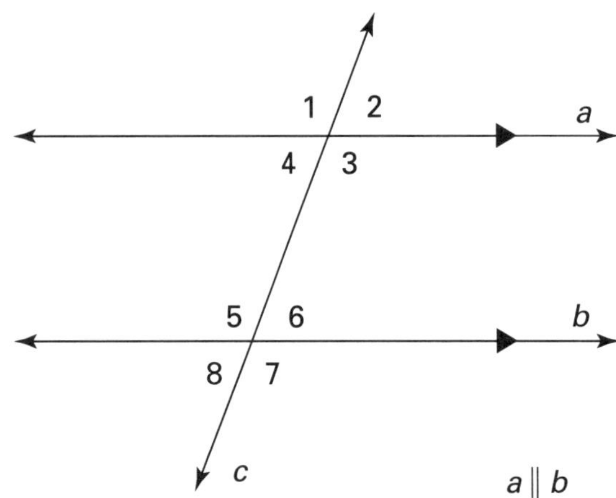

Step 1: Find an angle that is related to both angles. You know that ∠1 and ∠5 are congruent because they are corresponding angles. You also know that ∠5 and ∠6 are supplementary because they form a straight line (180°).

Step 2: Find the measure of the angle that has a relationship to both ∠1 and ∠6. You know that ∠1 and ∠5 are congruent because they are alternate exterior angles, and you know that ∠5 and ∠6 are supplementary angles (so, together, they add up to 180). Because you know that ∠1 is 115°, you also know that ∠5 is 115°.

Step 3: Find the measure of the unknown angle.

$\angle 5 + \angle 6 = 180°$

So, $\angle 6 = 180° - 115° = 65°$

Your final answer is 65°.

PRACTICE

18. Look at the figure. Then answer the following questions.

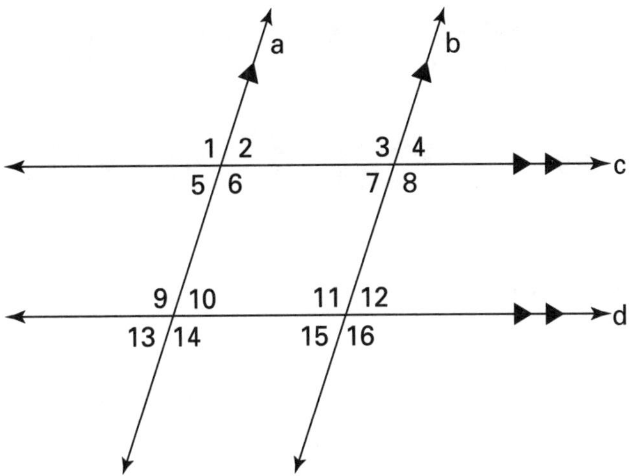

a. List the interior angles for each pair of parallel lines.
b. List the exterior angles for each pair of parallel lines.
c. Which angles correspond to $\angle 1$?
d. Which angle is an alternate interior angle with $\angle 5$?
e. Which angle is an alternate exterior angle with $\angle 5$?
f. List five pairs of vertical angles.
g. List ten sets of supplementary angles. Tell how you know each angle pair is supplementary.

19. Look at the figure again. If the measure of $\angle 12$ is $70°$, what is the measure of $\angle 16$?

20. Look at the figure once more. If the measure of $\angle 1$ is $120°$, what is the measure of $\angle 4$?

Finding Perimeter, Circumference, Area, and Volume

LESSON SUMMARY

In this lesson, you will learn how to measure different parts of geometric shapes. You'll learn how to find the perimeter, circumference, area, and volume of shapes.

Standardized tests often expect you to be able to measure different parts of geometric shapes. But this is not a skill you need to know only to pass a test. You need to know the perimeter of your yard if you want to enclose it with a fence. You need to know the area of your room if you want to carpet it. This lesson will not only help you pass any math test, it will also help you improve skills you will use for many different practical tasks.

FINDING PERIMETER AND CIRCUMFERENCE

Perimeter is the distance around a figure. You can calculate the perimeter of a figure by adding up the length of each side.

Example: Find the perimeter of a square whose side is 5 cm.

You know that each side of a square is equal. So, if you know that one side of a square is 5 cm, then you know that each side of the square is 5 cm. To calculate the perimeter, add up the length of all four sides.

$$5 + 5 + 5 + 5 = 20$$

So, the perimeter of a square whose side measures 5 cm is 20 cm.

Another way to calculate the perimeter of a square is $4 \times s$, where s = the length of one side of the square. So, the perimeter of a square whose side measures 5 cm would be $4 \times 5 =$ 20 cm.

Example: Find the perimeter of the following figure.

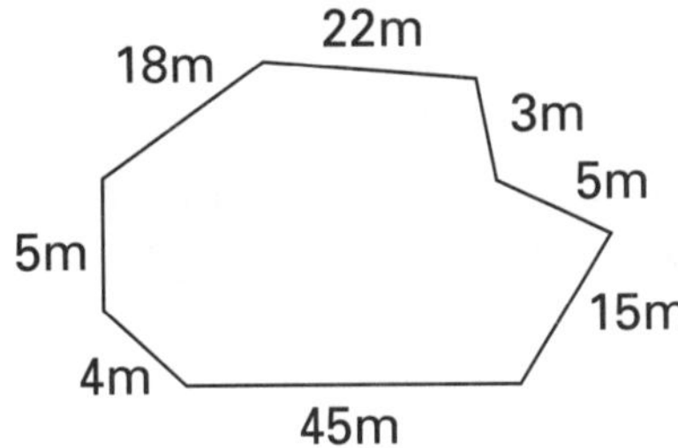

Add up the lengths of each side of the figure.

$$22 + 3 + 5 + 15 + 45 + 4 + 5 + 18 = 117$$

The perimeter of the figure is 117 meters.

When you are solving problems with units, be sure to always include the unit with your measurement. Numbers are not very useful or meaningful measurements unless you know the units in which your measurements are being made.

The perimeter of a circle is called its *circumference*. You can calculate a circle's circumference using either its radius or its diameter.

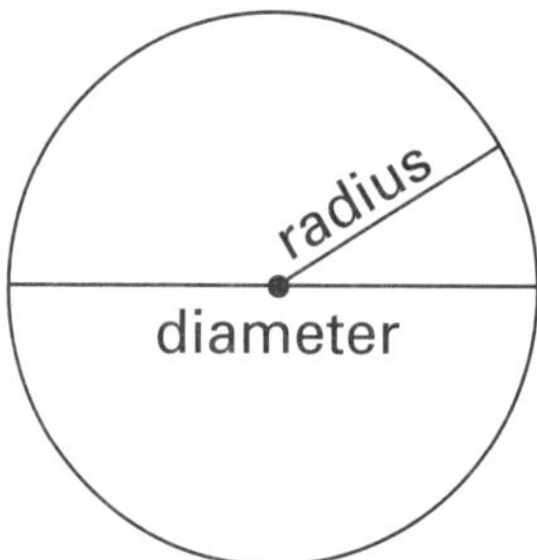

The *diameter* is a line segment that goes through the center of a circle. The endpoints of the diameter are on the circle. Any line that begins at the center of a circle and ends on a point on the circle is called a *radius*. A circle's diameter is twice as long as its radius. So, if you know either the radius or the diameter, you can easily find the other.

To calculate circumference, use either of these formulas, where $\pi = 3.14$ (π is a Greek symbol, spelled *pi*, and pronounced *pie*), d is the circle's diameter, and r is the circle's radius:

$C = \pi d$

$C = 2\pi r$

Example: Find the circumference of a circle with a diameter of 5 inches.

Since you know the diameter, use the formula that include the diameter: $C = \pi d$.

$$
\begin{aligned}
C &= \pi\,(5) \\
&= (3.14)(5) \\
&= 15.7
\end{aligned}
$$

The final answer is 15.7 inches.

If you're not allowed to use a calculator on a test, you can estimate the circumference of a circle by substituting 3 for π.

PRACTICE

Find the perimeter or circumference of each figure. You can check your answers at the end of the section.

1. a square with one side equal to 10 in

2. a rectangle with height equal to 3 ft and length equal to 5 ft

3. a triangle with sides 3 mm, 4 mm, and 5mm

4. a circle with radius 5 cm

5. a circle with diameter 10 m

6.

2ft 7ft 7ft 2ft

10ft 10ft

7. The measurements given below are in meters.

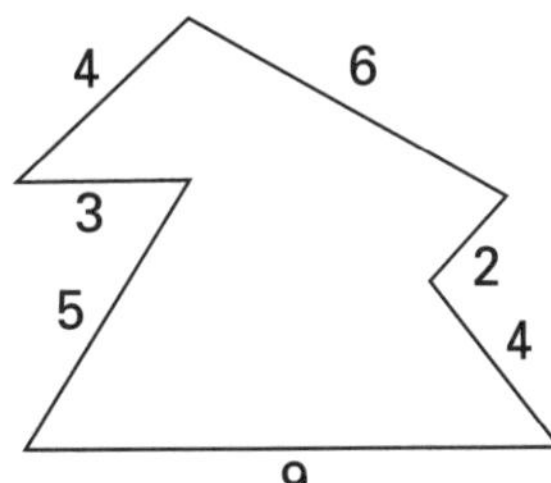

8. The figure below is exactly a quarter circle with a radius 8 cm.

Sometimes your answer may be given in terms of π. You treat the π symbol as an algebraic variable, and leave it in your answer. For example, if you were asked to find the circumference of a circle where $r = 2$, you would plug the 2 into your equation $2\pi r$: $2\pi(2) = 4\pi$. 4π would be your final answer.

FINDING AREA

Area is a measure of the surface of a two-dimensional figure. The following table shows how to calculate the area of different figures.

CALCULATING AREA

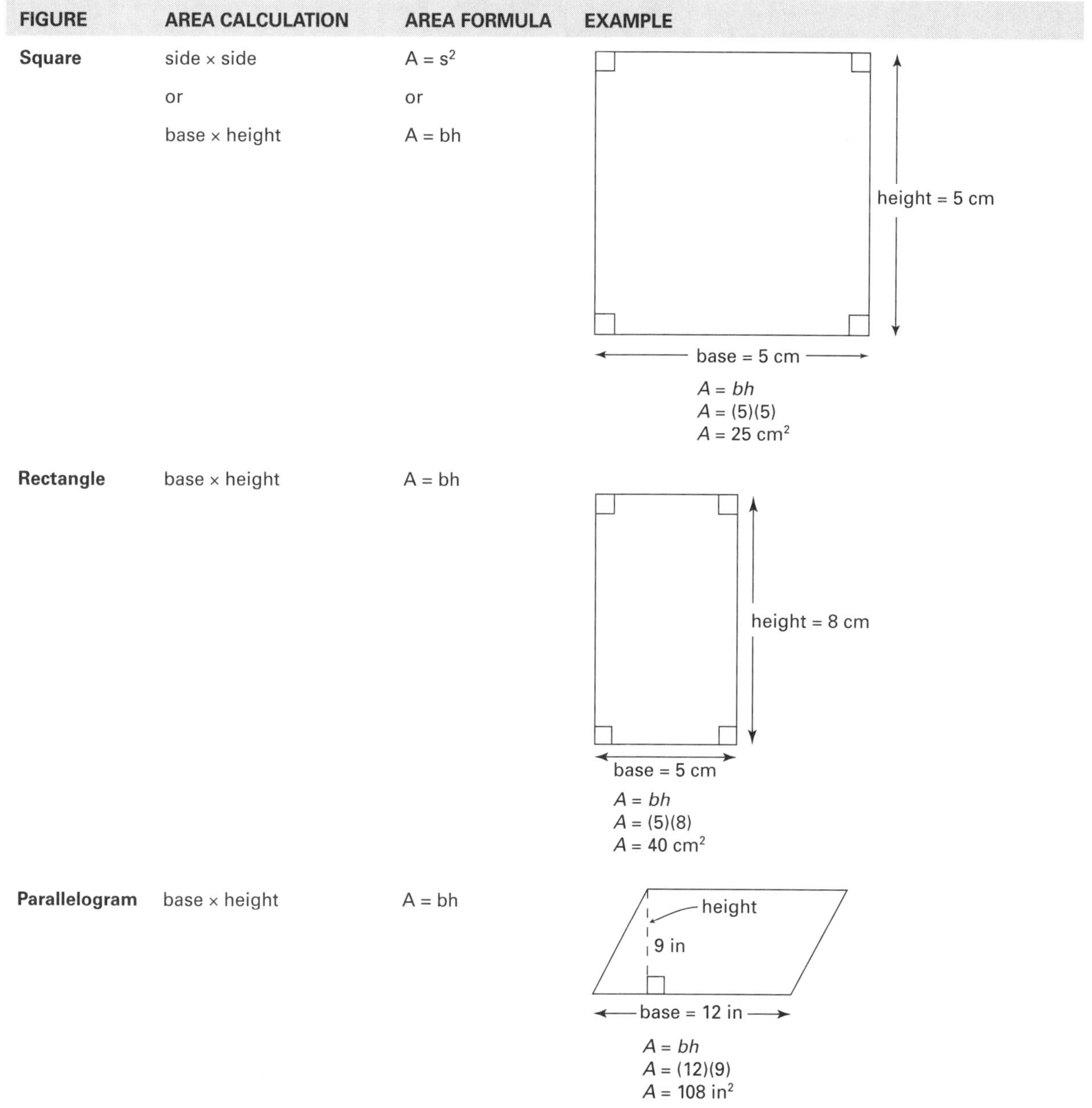

FIGURE	AREA CALCULATION	AREA FORMULA	EXAMPLE
Square	side × side or base × height	$A = s^2$ or $A = bh$	height = 5 cm base = 5 cm $A = bh$ $A = (5)(5)$ $A = 25\text{ cm}^2$
Rectangle	base × height	$A = bh$	height = 8 cm base = 5 cm $A = bh$ $A = (5)(8)$ $A = 40\text{ cm}^2$
Parallelogram	base × height	$A = bh$	height 9 in base = 12 in $A = bh$ $A = (12)(9)$ $A = 108\text{ in}^2$

FIGURE	AREA CALCULATION	AREA FORMULA	EXAMPLE
Triangle	$\frac{1}{2}$ base × height	$A = \frac{1}{2}bh$	height; 6 in; base = 10 in $A = \frac{1}{2}bh$ $A = \frac{1}{2}(10)(6)$ $A = 30 \text{ in}^2$
Trapezoid	$\frac{1}{2}$ × base_1 × height + $\frac{1}{2}$ × base_2 × height	$A = \frac{1}{2}h(b_1 + b_2)$	$\text{base}_1 = 7$ cm; height = 6 cm; $\text{base}_2 = 17$ cm $A = \frac{1}{2}h(b_1 + b_2)$ $A = \frac{1}{2}(6)(7 + 17)$ $A = \frac{1}{2}(6)(24)$ $A = 72 \text{ cm}^2$
Circle	π × radius squared	$A = \pi r^2$	diameter = 20 ft $A = \pi r^2$ $r = \frac{1}{2}d$ $r = \frac{1}{2}(20) = 10$ feet $A = 3.14 \times 10 \times 10$ $A = 314 \text{ ft}^2$

A figure's base is the side that forms a right angle with the height of the figure. The base does not have to be the bottom side of a figure. Notice that a trapezoid has both a top and a bottom base, for example.

Example: Find the area of the figure.

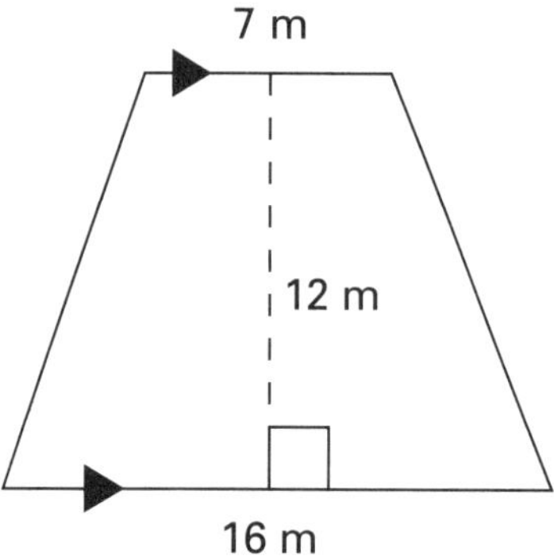

Step 1: Choose the correct formula.

From the table you know that this figure is a trapezoid. So, you need to use the following formula: $A = \frac{1}{2}h(b_1 + b_2)$.

Step 2: Plug in the measures for the variables in the formula and solve.

$A = \frac{1}{2}h(b_1 + b_2)$

$A = \frac{1}{2}(12)(7 + 16)$

$A = \frac{1}{2}(12)(23)$

$A = \frac{1}{2}(276)$

$A = 138$

So the final answer is 138 m^2.

Notice that to calculate the area of a figure, you need two measurements. Multiplying the units of each measurement together gives you *square units*. That's why area is measured in square units. (Perimeter was measured in *linear* units.)

Examples of square units are:

square miles or mi^2
square meters or m^2
square centimeters or cm^2
square feet or ft^2
square inches or in^2

Example: The figure below has an area of 40 cm^2. Find the height of the figure.

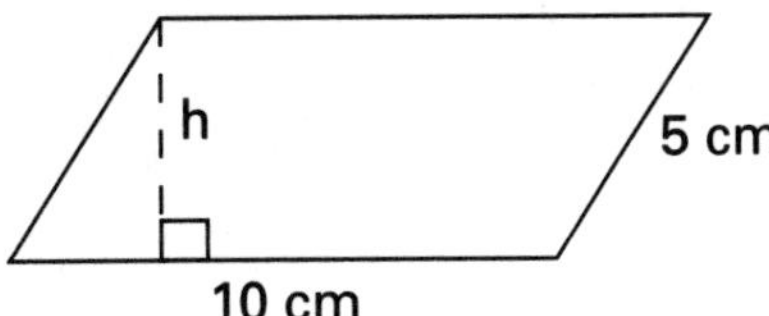

Step 1: Choose the correct formula.

From the table you know that this figure is a parallelogram. So, you need to use the following formula: $A = bh$.

Step 2: Plug in the known measures for the variables in the formula and solve for the height.

$A = bh$

$40 = 10h$

$\frac{40}{10} = \frac{10h}{10}$

$4 = h$

So the final answer is 4 cm.

PRACTICE

Find the area of each figure. You can check your answers at the end of the section.

9.

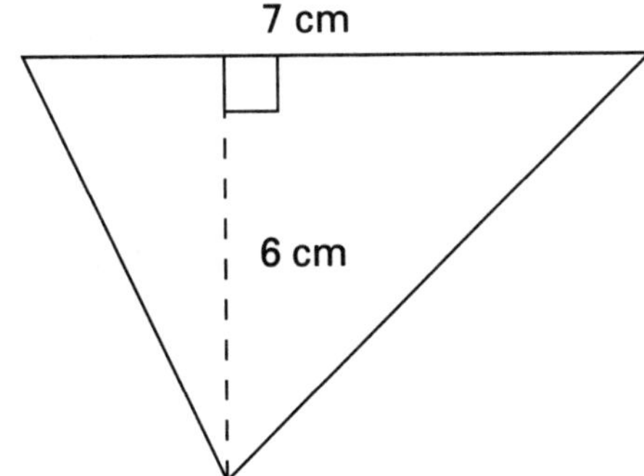

10.

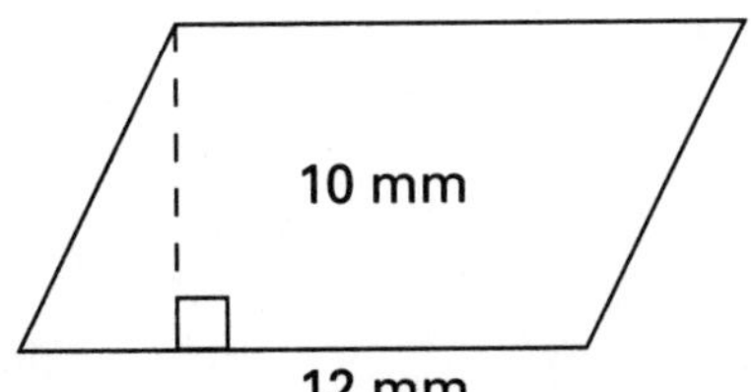

11.

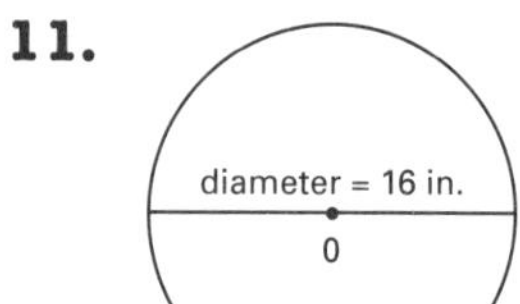

Solve each problem. Check your answers at the end of the section.

12. Find the measure of b_1.

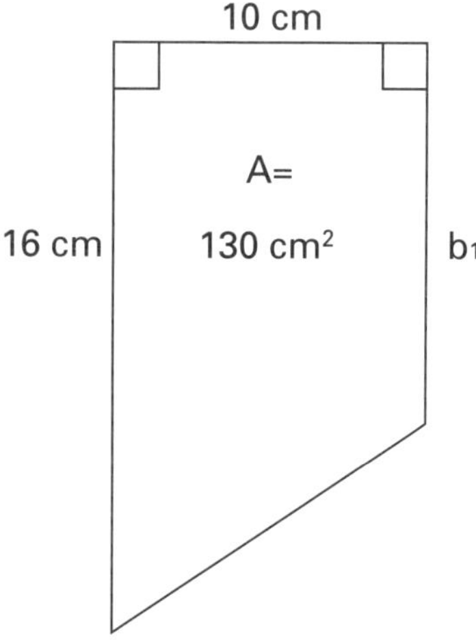

13. How much greater is the area of circle A than the area of circle B?

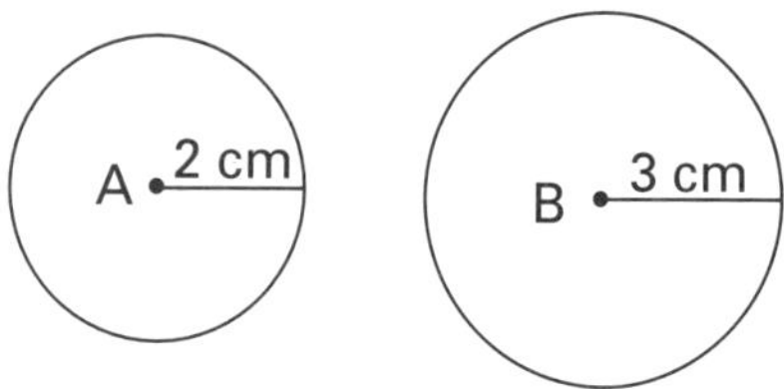

14. Find the area of the shaded portion in the figure below. (Hint: the sides of the square are equal to the diameter of the circle.)

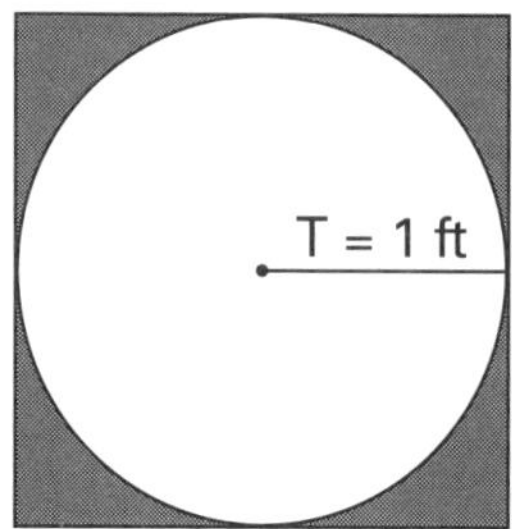

15. The area of the circle shown below is 100π cm^2. What is the length of side AB? (Hint: Don't forget that $2r = d$)

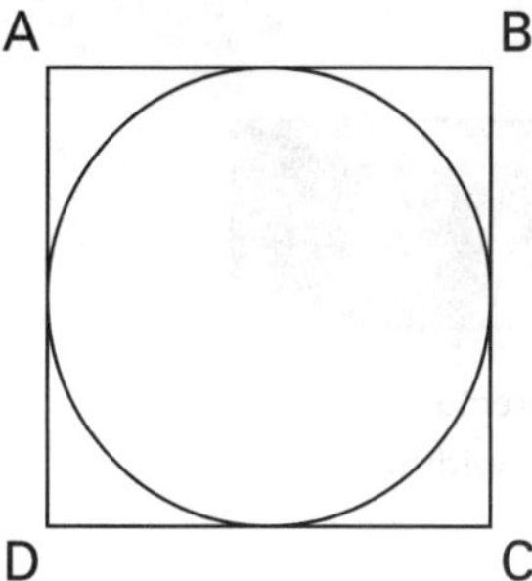

FINDING VOLUME

Volume is a measure of the amount of space inside a three-dimensional shape. Three-dimensional shapes are sometimes called solids. Examples of three-dimensional shapes are shown below.

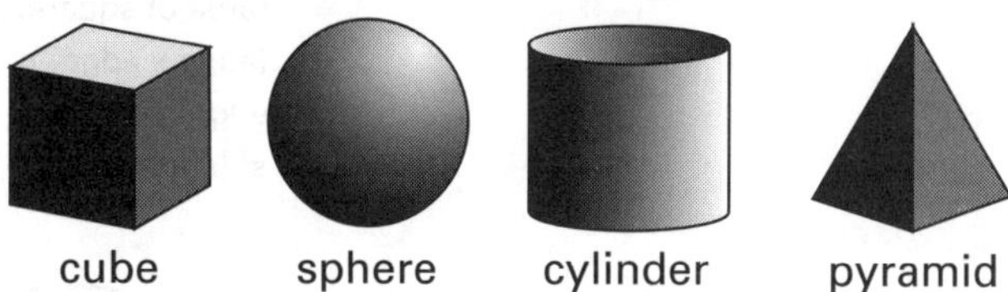

In this lesson, you will learn how to calculate the volume of a rectangular solid, a cube, and a cylinder. The formula for calculating the area of these solids is the same.

$V = Ah$
Where V is the volume
A is the area of the base
h is the height

Let's see how it works.

FIGURE	VOLUME FORMULA	EXAMPLE
Rectangular solid	$V = Ah$ $V = lwh$	V = (area of rectangle) h V = (length x width) x height = lwh
Cube	$V = Ah$ $V = s^3$	V = (area of square) h V = (edge x edge) x edge = s^3 (edge = side)
Square	$V = Ah$ $V = \pi r^2 h$	 V = (area of circle) h V = (π x radius^2) x height = $\pi r^2 h$

Notice that to calculate the volume of a figure, you need three measurements. Multiplying the units of each measurement together gives you *cubic units*. You can imagine a cubic unit as a cube with one unit edges.

Example: Find the volume of a cube that is 3 inches long on each edge.

Step 1: Choose the correct formula. The problem tells you that you are measuring the volume of a cube. You know from the table that the formula for the volume of a cube is $V = Ah$ or $V = s^3$.

Step 2: Plug in the known measures and solve.

$$V = 3^3$$
$$V = 3 \times 3 \times 3$$
$$V = 27$$

So the final answer is 27 in^3.

Example: Find the volume of a cylinder that has a height of 10 cm and a radius of 5 cm.

Step 1: Choose the correct formula. The problem tells you that you are measuring the volume of a cylinder. You know from the table that the formula for the volume of a cylinder is $V = Ah$ or $V = \pi r^2 h$.

Step 2: Plug in the known measures and solve.

$$V = \pi r^2 h$$
$$V = \pi(5)^2(10)$$
$$V = \pi(25)(10)$$
$$V = \pi(250)$$
$$V = 785$$

So the final answer is 785 cm^3.

PRACTICE

Find the volume of each figure.

16.

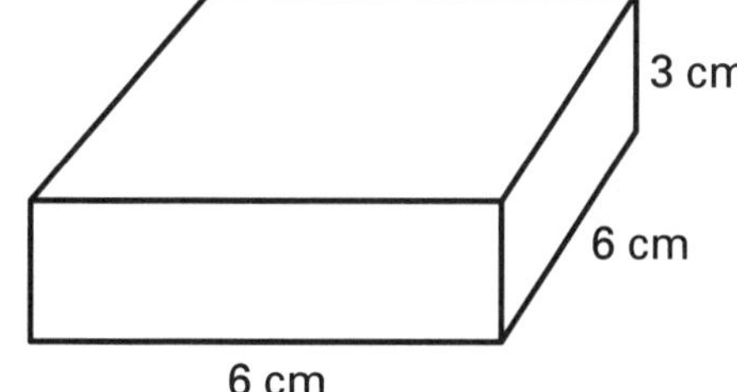

17.

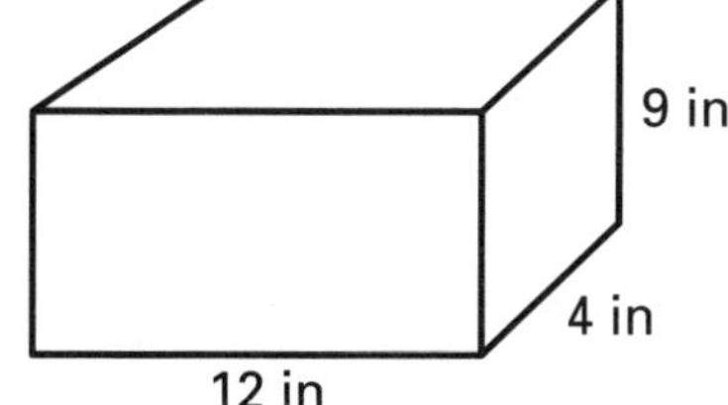

18.

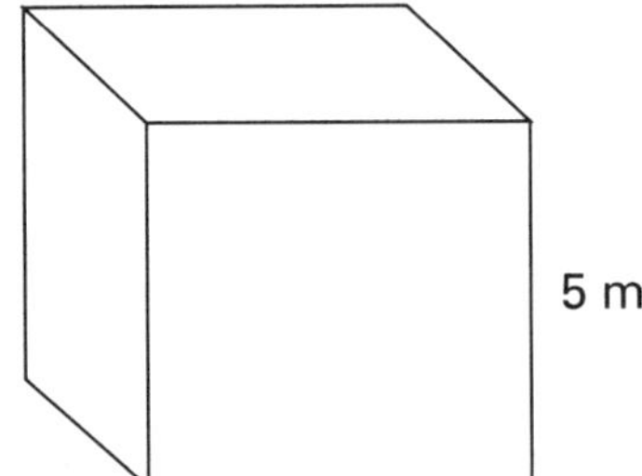

19.

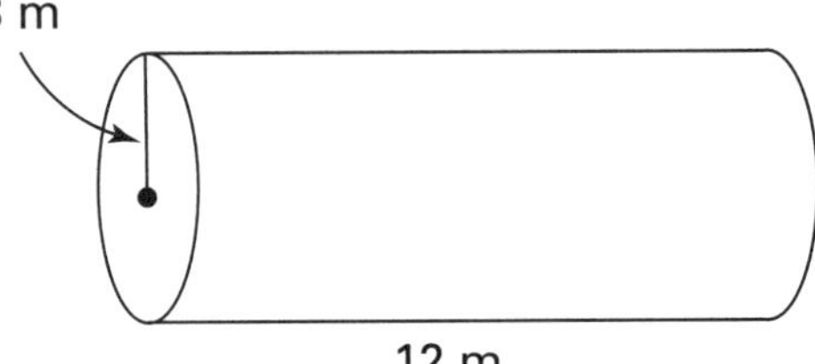

20. The shaded portion of the cylinder is filled with water. If the water is 4 cm high, what is the volume of the water in the cylinder?

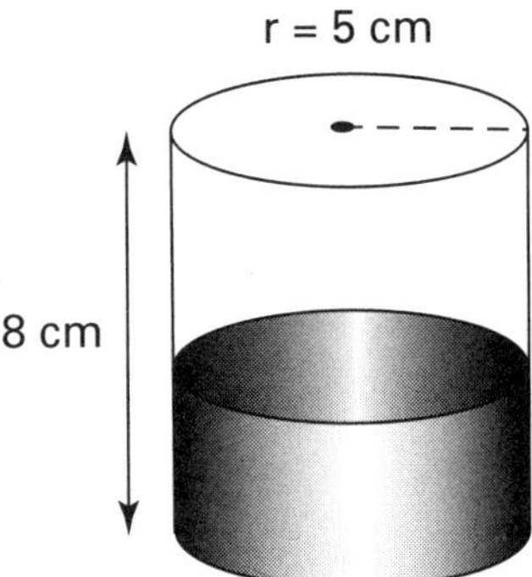

LESSON

20

Congruence and Similarity

LESSON SUMMARY

This lesson introduces the concepts of congruence and similarity. You will learn how to tell if two triangles are congruent or if they are similar. You will also learn how to use these concepts to solve basic geometry problems.

Two figures that are exactly the same shape and the same size are said to be *congruent*—they are exactly identical to one another. Congruent figures are often used in constructing buildings, quilt patterns, and other things. In contrast, *similar* figures are the same shape, but they are not always the same size. Instead, their sides are in proportion to one another. You can use similar figures to determine measures that would otherwise be very difficult to measure—such as the height of a tall building. In this lesson, you'll learn about congruent and similar triangles.

CONGRUENT TRIANGLES

Triangles are *congruent* if they are exactly the same size and the same shape. You might look at two triangles and guess whether they are the same—you're probably a pretty good judge as to whether two triangles are the same size and shape. Or, you could cut out one of the triangles and see if it fits exactly on top of the other one. If so, they are congruent.

However, geometry is largely about *proving* things, not guessing about them. So, you need to be able to use basic geometry rules, such as *theorems* (formulas or statements in mathematics that can be proved true) to show that two triangles are congruent.

Though geometry rules come in different forms, they can be generally understood as statements that all mathematicians have agreed to accept as true or that can be proved true. *Postulates* are statements that are accepted without proof. *Theorems* are statements that can be proved true. You will learn more about these principles when you take a geometry course. For now, you just need to know that you must follow certain steps to prove that two triangles are congruent.

In fact, there are three rules for proving that two triangles are congruent to one another. Notice that the symbol ≅ means congruent. When you see this symbol, say the word "congruent."

PROVING THAT TRIANGLES ARE CONGRUENT

RULE NAME	WHAT IT SAYS	WHAT IT LOOKS LIKE
Side-Side-Side (SSS)	If three sides of one triangle are congruent to three sides of another triangle, then the two triangles are congruent. (Note: The lines on each side of both triangles corresponds with the side of the compared triangle that is equal.)	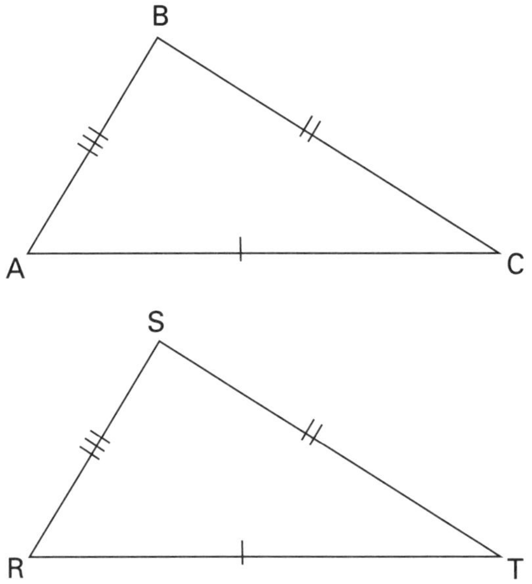 $\Delta ABC = \Delta RST$

RULE NAME	WHAT IT SAYS	WHAT IT LOOKS LIKE
Side-Angle-Side (SAS)	If two sides and the angle they form are congruent to the corresponding parts of another triangle, then the two triangles are congruent.	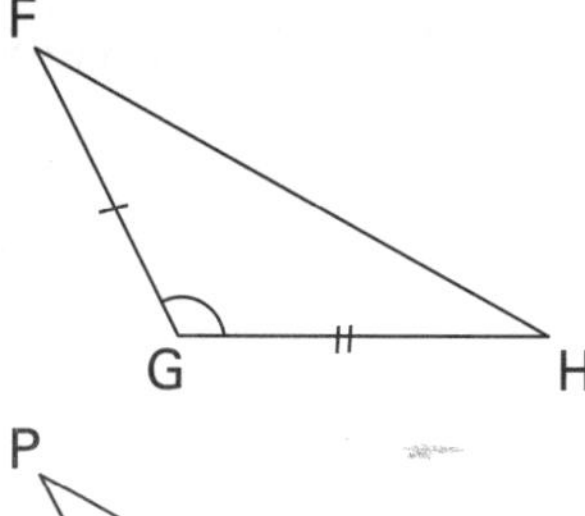 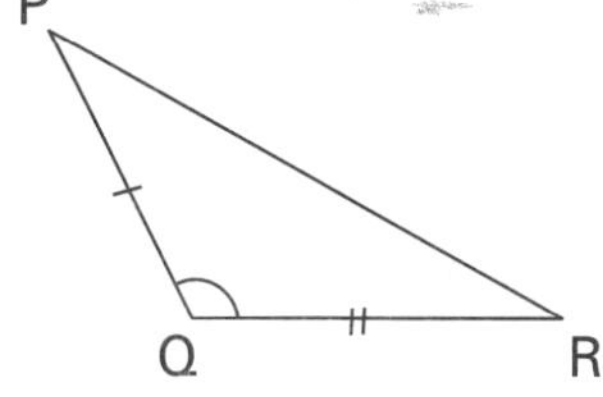 ΔFGH = ΔPQR
Angle-Side-Angle (ASA)	If two angles and the side in between them of one triangle are congruent to the corresponding parts of another triangle, then the two triangles are congruent.	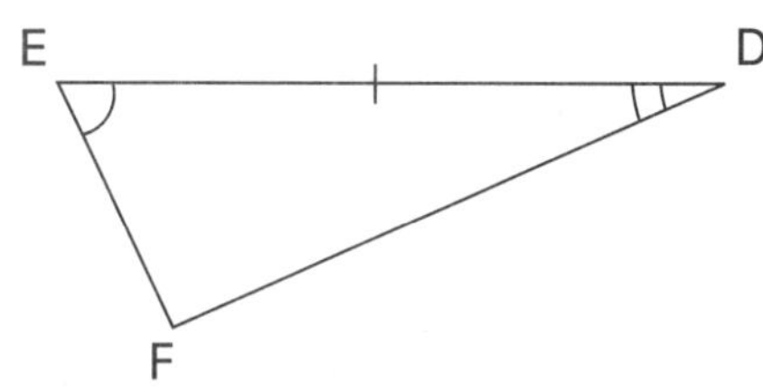 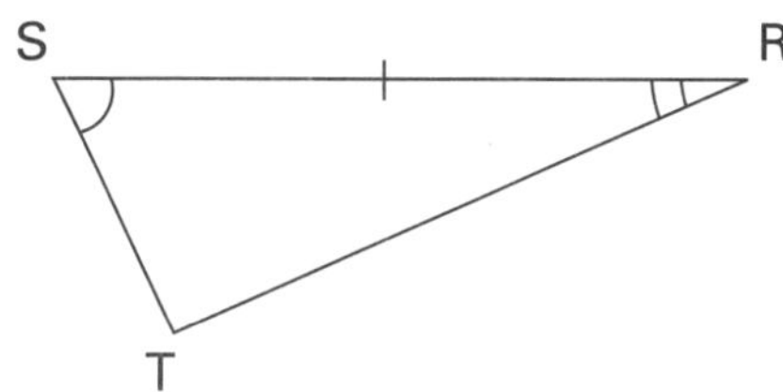 ΔDEF = ΔRST

What does *corresponding part* mean? This term refers to the part that is being compared in both triangles. So, if you are comparing two triangles, you'll want to name the parts in the same order in each triangle. You'll want to compare the hypotenuse (the side of a right triangle that is opposite the right angle) of one right triangle with the hypotenuse of another triangle. You'll want to compare the 75° angle of one triangle with the 75° angle of the other triangle. Notice in the table on pages 308–309 that the parts with the same relative position (the same number of marks on the side) in each triangle are always compared.

Here's how it works.

Example: Is ∠ABC ≅ ∠FED? (The ≅ symbol means congruent.)

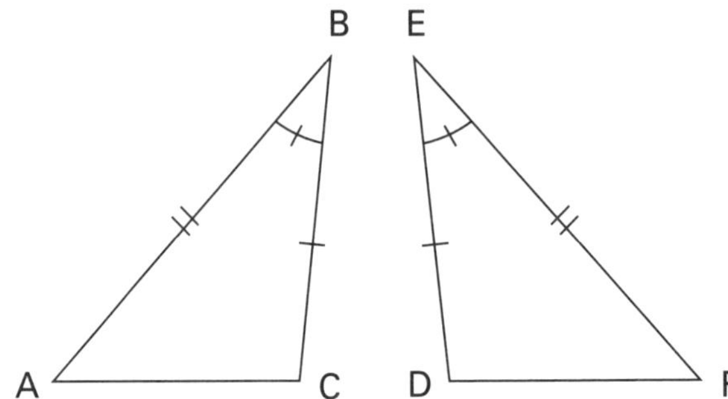

Step 1: Ask yourself: which corresponding parts of the triangles does the problem tell me are congruent? You know from the symbols on the triangles that

$\overline{AB} \cong \overline{FE}$
$\overline{BC} \cong \overline{ED}$
$\angle B \cong \angle E$

Step 2: Ask yourself: Is one of the rules needed to prove that two triangles are congruent met? Yes, the given information matches the rule for Side-Angle-Side.
Step 3: Answer the question. Since the criteria for Side-Angle-Side are met, the two triangles are congruent.

Therefore, the two triangles are congruent: ΔABC ≅ ΔFED.

Example: Is ΔPQR ≅ ΔZYX?

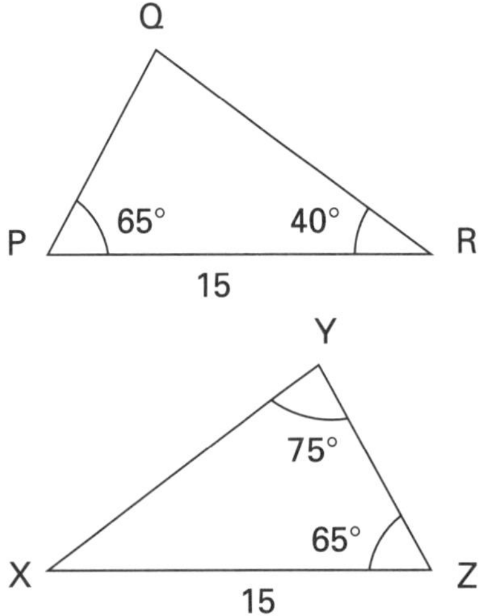

Step 1: Ask yourself: which corresponding parts of the triangles does the problem tell me are congruent? You know from the symbols on the triangles that $\overline{PR} \cong \overline{ZX}$ and ∠P ≅ ∠Z.

Step 2: Ask yourself: Is one of the rules needed to prove that two triangles are congruent met? Not yet. You know only one side and one angle so far.

Step 3: Figure out what information you still need to know to make a decision. You need to know the measure of ∠X. You know that the sum of the angles in a triangle is 180°. So you can determine the measure of ∠X as follows:

$$180° - 75° - 65° = 40°$$

The measure of ∠X is 40°.

Step 4: Answer the question. Now you know that Angle-Side-Angle is true, so the two triangles are congruent.

ΔPQR ≅ ΔZYX.

PRACTICE

Prove that each pair of triangles is congruent.

1.

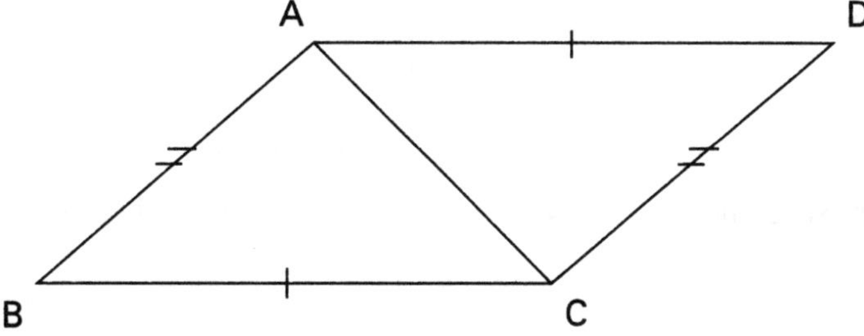

2.

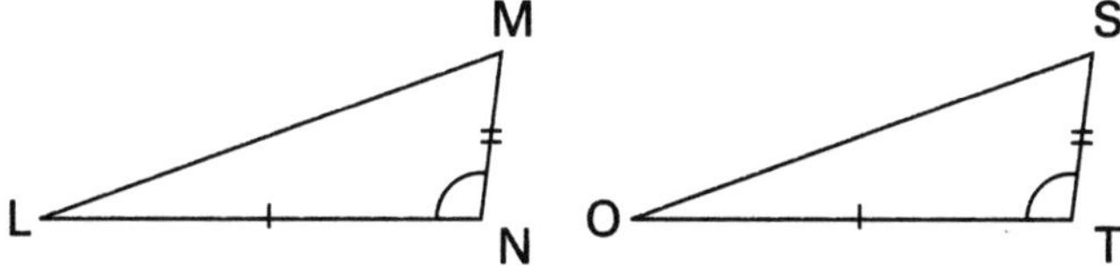

3.

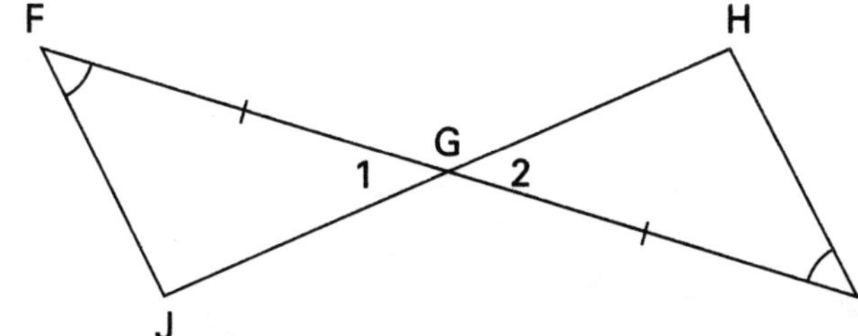

4.

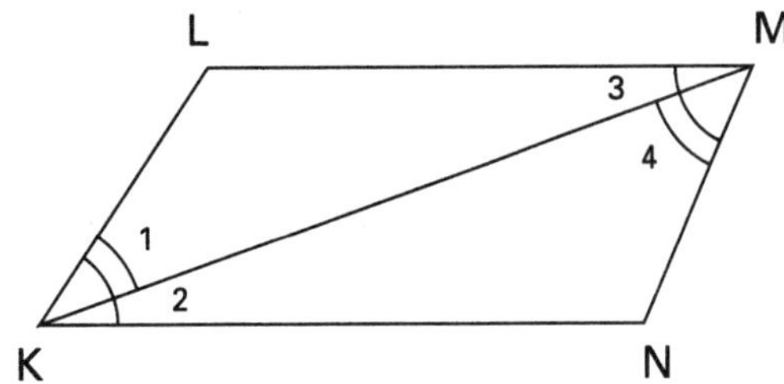

5.

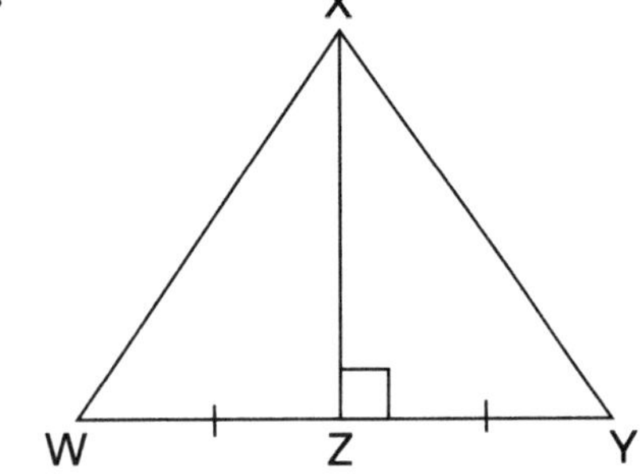

SIMILAR TRIANGLES

Triangles are *similar* if they have the same shape and their sizes are proportional to one another. You can prove that two triangles are similar using one of the three rules listed on the following table. Notice that the symbol ~ means similar.

PROVING THAT TRIANGLES ARE SIMILAR

RULE NAME	WHAT IT SAYS	WHAT IT LOOKS LIKE
Angle-Angle (AA)	If two angles of one triangle are congruent to two angles of another triangle, then the two triangles are similar.	$\angle A \cong \angle D \quad \angle C \cong \angle F$ $\Delta ABC \sim \Delta DEF$
Side-Side-Side (SSS)	If the lengths of the corresponding sides of two triangles are proportional, then the two triangles are similar	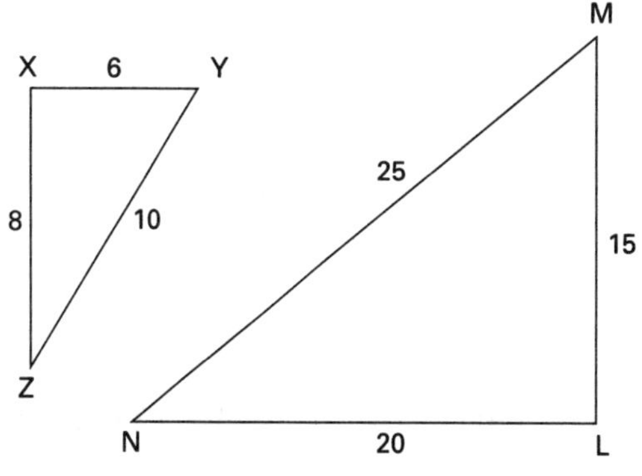 $\frac{6}{15} \stackrel{?}{=} \frac{10}{25}$ $\frac{6}{15} \stackrel{?}{=} \frac{8}{20}$ $\frac{8}{20} \stackrel{?}{=} \frac{10}{25}$ $6 \times 25 = 15 \times 10$ $6 \times 20 = 15 \times 8$ $8 \times 25 = 20 \times 10$ $150 = 150$ $120 = 120$ $200 = 200$ $\Delta XYZ \sim \Delta LMN$

RULE NAME	WHAT IT SAYS	WHAT IT LOOKS LIKE
Side-Angle-Side (SAS)	If the lengths of two pairs of corresponding sides of two triangles are proportional and the corresponding angles in between those two sides are congruent, then the two triangles are similar.	K, 12, S, 4, L, 16, J, R, T $\frac{3}{4} \stackrel{?}{=} \frac{12}{16}$ 3 × 16 = 4 × 12 48 = 48 ΔRST ~ ΔJKL

Example: Is ∠MNO ~ ∠JKL? (Remember, the ~ symbol means similar.)

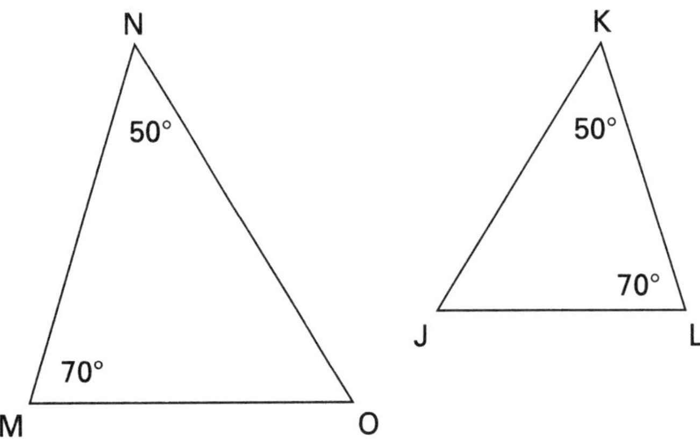

Step 1: Ask yourself: which corresponding parts of the triangles are likely to be similar? You know from the symbols on the triangles that

∠N = ∠K
∠M = ∠L

Step 2: Ask yourself: Is one of the rules needed to prove that two triangles are similar met? Yes, two angles are congruent.

Step 3: Answer the question. Since Angle-Angle is true, the two triangles are similar.

ΔMNO ~ ΔLKJ.

If two triangles you are comparing are turned in different directions and it's hard to keep track of which parts correspond, quickly redraw one of the triangles so it's oriented the same way as the other one. For example, in the last example, you might have redrawn the second triangle as shown below to make the problem easier.

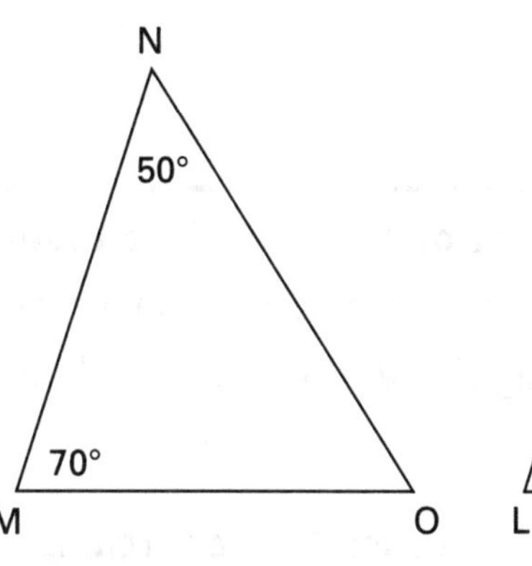

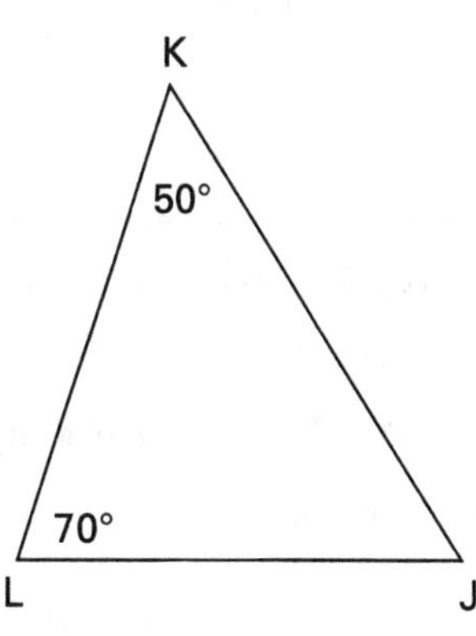

Example: Is ΔABC ~ ΔXYZ?

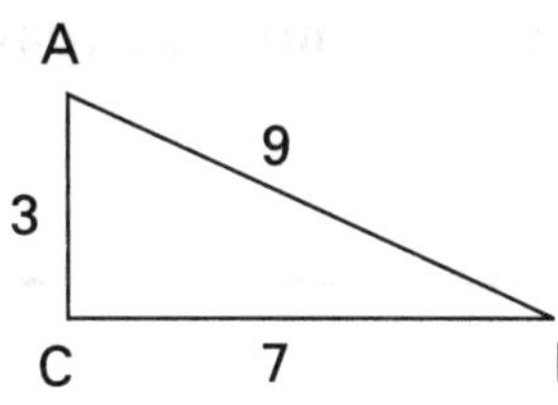

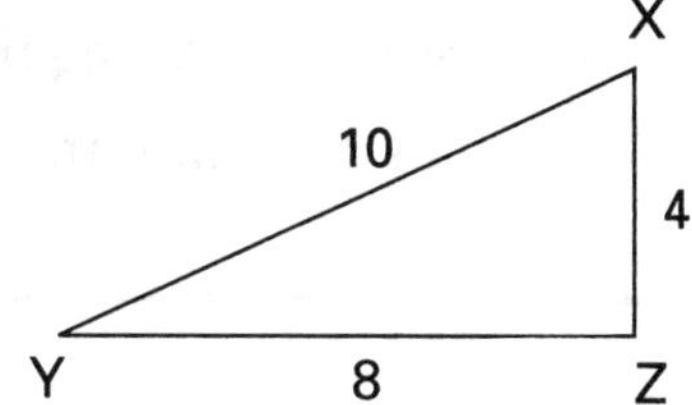

Step 1: Ask yourself: which corresponding parts of the triangles are likely to be similar? You know from the shapes of the triangles that

$\overline{AB}$ corresponds with $\overline{XY}$
$\overline{BC}$ corresponds with $\overline{YZ}$
$\overline{AC}$ corresponds with $\overline{XZ}$

Step 2: Ask yourself: Is one of the rules needed to prove that two triangles are similar met? Since you do not know the measures of the any of the angles of either triangle, you will have to compare the sides of the triangles. To find out if the sides of the triangles are in proportion, you need to set up a proportion and see if it is true, as shown below.

$\frac{3}{4} \stackrel{?}{=} \frac{9}{10}$

To test whether the proportion is true, you can cross multiply:

$3 \times 10 \stackrel{?}{=} 4 \times 9$
$30 \neq 36$

Because the two sides of the equation are not equal, the sides are not proportional. Side-Side-Side is not true.

Step 3: Answer the question. Since none of the rules that prove two triangles are similar are met, the two triangles are not similar.

ΔABC and ΔXYZ are not similar triangles.

Another way to tell if a proportion is correct is to reduce both ratios to lowest terms. If they reduce to the same ratio, then they are equal. Let's look again at the last problem to see how this works.

You started with the following possible proportion.

$\frac{3}{4} \stackrel{?}{=} \frac{9}{10}$

You know that $\frac{3}{4}$ is already reduced to lowest terms. Can $\frac{9}{10}$ be reduced to lower terms? No. Since $\frac{3}{4} \neq \frac{9}{10}$, you know that the proportion is not true. So again, you can conclude that the two triangles are not similar.

Example: At 3 P.M. a five-meter pole makes a shadow that is 3 meters long. At the same hour, a tree makes a 21-meter shadow. How tall is the tree? Assume that the two triangles are similar.

Step 1: You know from the problem that the triangles are similar. So you can set up a proportion to solve for the missing length (the height of the tree).

$$\frac{\text{Pole's height}}{\text{Tree's height}} = \frac{\text{Length of the pole's shadow}}{\text{Length of the tree's shadow}}$$

Step 2: You are looking for the height of the tree, so let x = the height of the tree. Then, plug in the measures and solve for x.

$$\frac{5}{x} = \frac{3}{21}$$

Cross multiply to solve for x.

$5 \times 21 = 3x$

$105 = 3x$

$\frac{105}{3} = \frac{3x}{3}$

$35 = x$

So the tree is 35 meters tall.

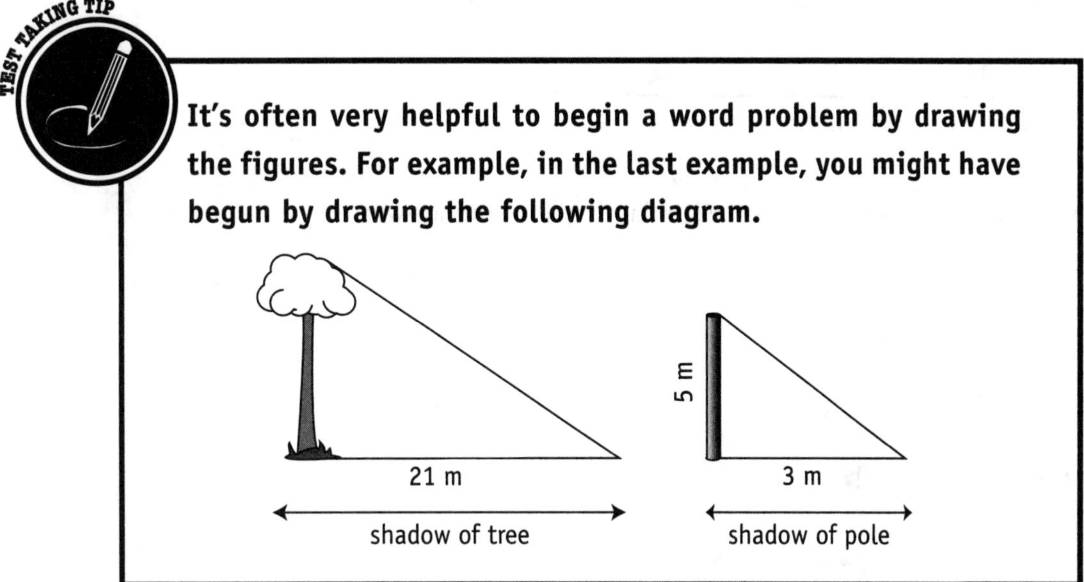

PRACTICE

Solve each problem using the principles of similarity. You can check your answers at the end of the section.

6. Tell whether these two triangles are similar. Explain your answer.

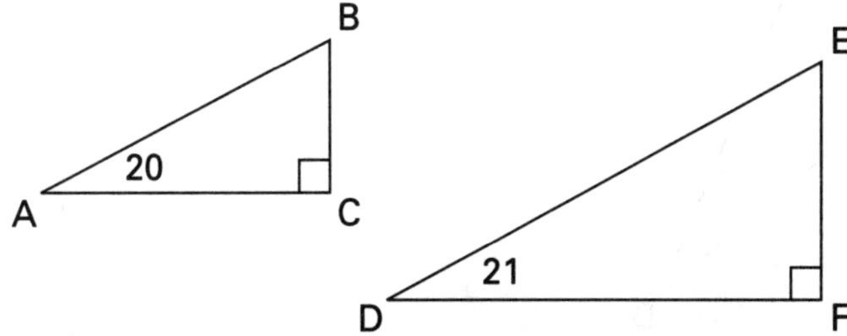

7. Tell whether these two triangles are similar. Explain your answer.

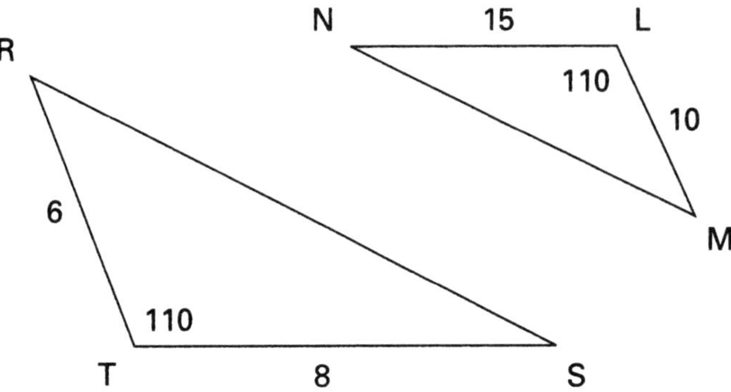

8. Tell whether these two triangles are similar. Explain your answer.

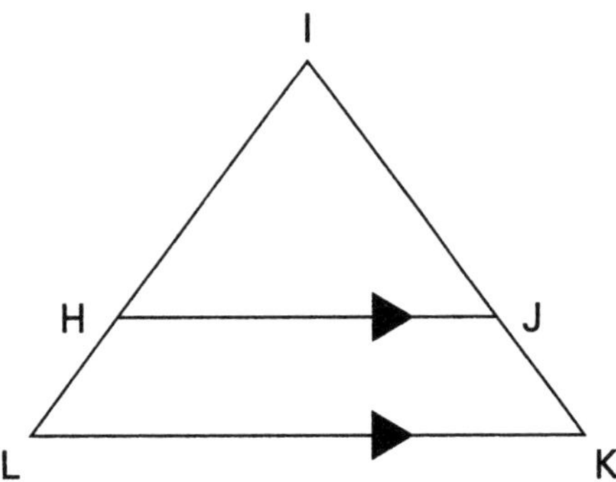

9. Tell whether these two triangles are similar. Explain your answer.

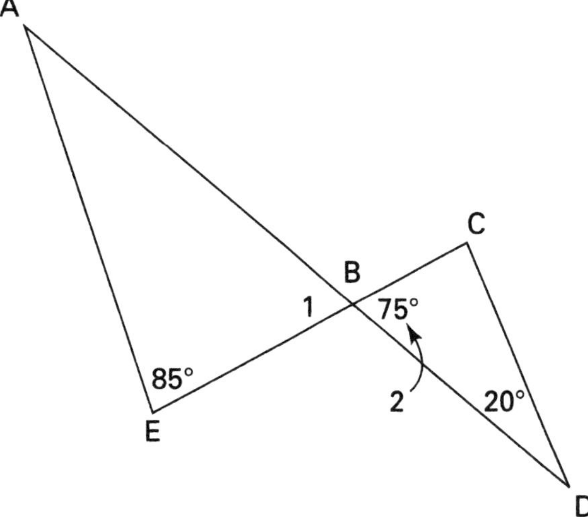

10. At 2 P.M. a flagpole makes a shadow that is 20 feet long. At the same hour, a 6-foot tall person makes a 4-foot long shadow. How tall is the flagpole? Assume that the two triangles are similar.

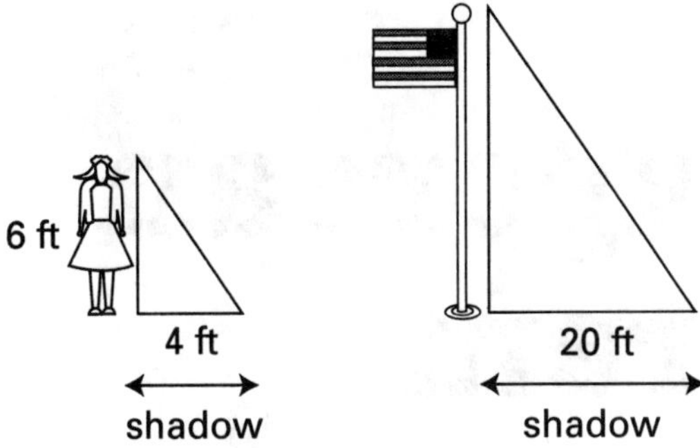

The Pythagorean Theorem

LESSON SUMMARY

This lesson introduces the parts of a right triangle. You will also learn what the Pythagorean Theorem is and how to use it to calculate the unknown lengths of the sides of right triangles.

You know that a triangle is a figure made up of three sides that come together to form three angles. In Lesson 18, you learned that an angle measuring exactly 90° is called a right angle. A triangle that has a right angle is called a *right triangle.* The sides and angles of right triangles have special relationships. In this lesson, you will learn about one of those relationships, which is called the Pythagorean Theorem.

THE PARTS OF A RIGHT TRIANGLE

Before you can use the Pythagorean Theorem, you need to know the basic parts of a right triangle. As you know, a right triangle has three sides. Two of the sides come together to form the right angle. These two sides are called *legs.* The third side of the triangle is called the *hypotenuse.* The hypotenuse is always the longest side of a right triangle. It is directly across from the right angle. Here are some examples of right triangles.

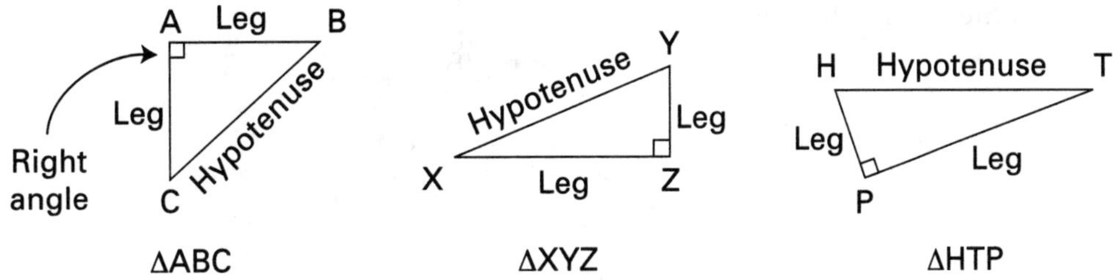

Example: Name the legs and the hypotenuse of ΔABC.

A right triangle has two legs. In ΔABC, the legs are $\overline{AB}$ and $\overline{AC}$. The hypotenuse is the longest side and is directly across from the right angle. In ΔABC, the hypotenuse is $\overline{CB}$.

PRACTICE

Use the right triangles above to answer the following questions.

1. Name the hypotenuse in ΔXYZ.
2. Name the legs of ΔHTP.
3. Name the legs of ΔXYZ.
4. Name the hypotenuse in ΔHTP.

▶ THE PYTHAGOREAN THEOREM

The equation below summarizes the Pythagorean Theorem.

$$a^2 + b^2 = c^2$$

In the equation, a and b are the legs of the right triangle, and c is the hypotenuse. In words, the Pythagorean Theorem says the following:

In a right triangle, the sum of the squares of the lengths of the legs is equal to the square of the length of the hypotenuse.

Let's see how this works in a few examples.

Example: Find the missing length.

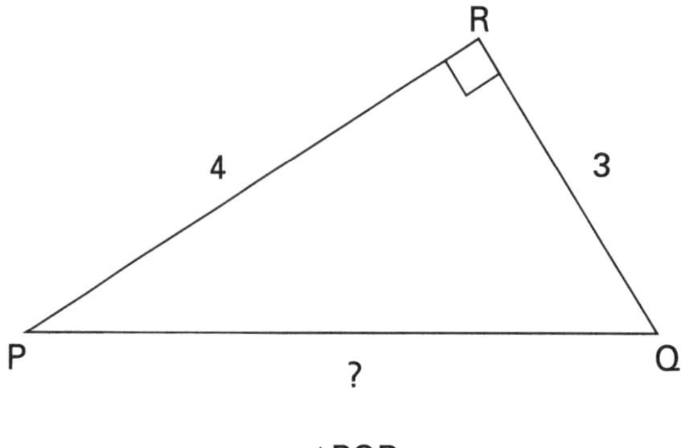

ΔPQR

Step 1: Find the legs and the hypotenuse in the right triangle.

The legs are $\overline{PR}$ and $\overline{RQ}$. The hypotenuse is $\overline{PQ}$.

Step 2: Match up the variables a, b, and c with the parts of the right triangle.

Let a = the length of $\overline{PR}$.

Let b = the length of $\overline{RQ}$.

Let c = the length of $\overline{PQ}$.

Step 3: Plug the known lengths into the Pythagorean Theorem.

$a^2 + b^2 = c^2$

$4^2 + 3^2 = c^2$

Step 4: Solve.

$4^2 + 3^2 = c^2$

$16 + 9 = c^2$

$25 = c^2$

$\sqrt{25} = \sqrt{c^2}$

$5 = c$

So, the length of the hypotenuse is 5.

It's a good idea to know basic squares and square roots by memory. Here are some common squares and square roots you might want to learn. You can review squares and square roots in Lesson 2.

continued from previous page

Squares to Know

Number	Square
1	1
2	4
3	9
4	16
5	25
6	36
7	49
8	64
9	81
10	100
11	121
12	144
13	169
14	196
15	225
16	256
17	289
18	324
19	361
20	400
21	441
22	484
23	529
24	576
25	625

Square Roots to Know

Number	Square Root
1	1
4	2
9	3
16	4
25	5
36	6
49	7
64	8
81	9
100	10
121	11
144	12
169	13
196	14
225	15
256	16
289	17
324	18
361	19
400	20
441	21
484	22
529	23
576	24
625	25

Example: Find the missing length.

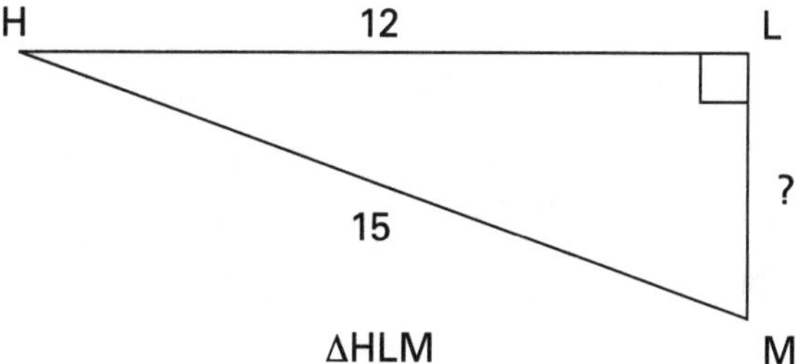

Step 1: Find the legs and the hypotenuse in the right triangle. The legs are $\overline{HL}$ and $\overline{LM}$. The hypotenuse is $\overline{HM}$.

Step 2: Match up the variables a, b, and c with the parts of the right triangle.

Let a = the length of $\overline{HL}$.

Let b = the length of $\overline{LM}$.

Let c = the length of $\overline{HM}$.

Step 3: Plug the known lengths into the Pythagorean Theorem.

$$a^2 + b^2 = c^2$$
$$12^2 + b^2 = 15^2$$

Step 4: Solve.

$$\begin{aligned} 12^2 + b^2 &= 15^2 \\ 144 + b^2 &= 225 \\ 144 + b^2 &= 225 \\ -144 \quad & \quad -144 \\ b^2 &= 81 \\ \sqrt{b^2} &= \sqrt{81} \\ b &= 9 \end{aligned}$$

So the length of the missing leg is 9.

You can also use the Pythagorean Theorem to determine if a triangle is a right triangle or not. Plug in the lengths of the sides of a triangle. If $a^2 + b^2 = c^2$ is true for that triangle, then you know it's a right triangle. You must know the length of all three sides for this to work. You can also memorize some common, small numbered groups of triples (such as (3, 4, 5); (5, 12, 13), and (9, 12, 15)) to help you solve problems more quickly and easily.

PRACTICE

Find each missing length.

5.

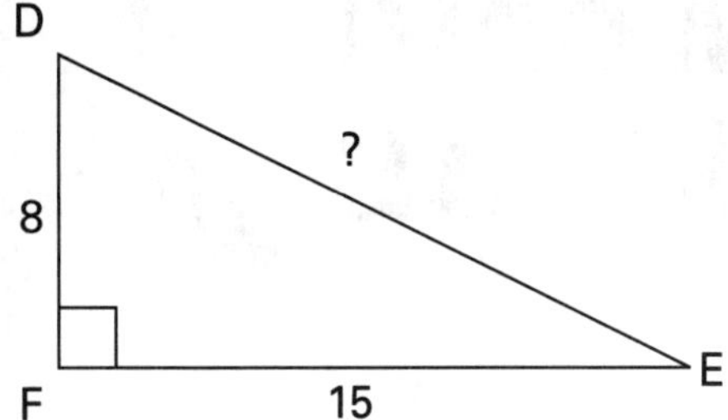

6.

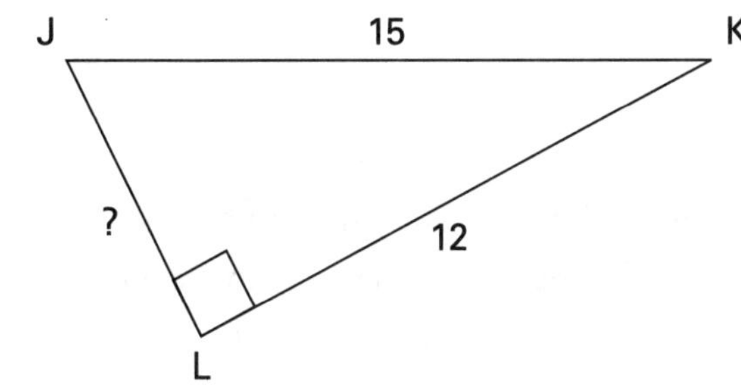

7.

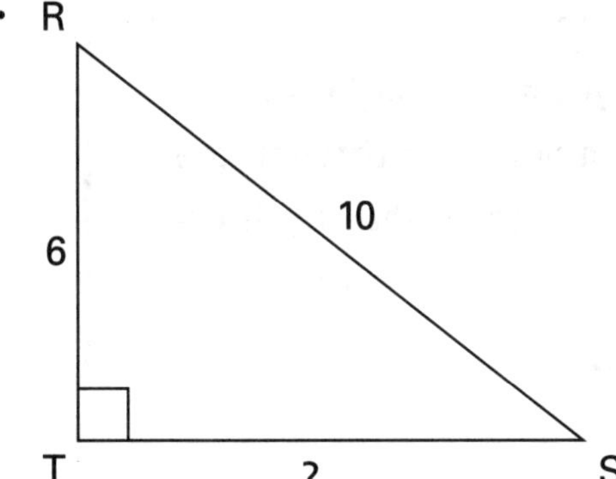

Determine if each triangle is a right triangle.

8. Let $a = 5$
Let $b = 6$
Let $c = 9$

9. A triangle with sides 6, 8, and 10

10. A triangle with sides 3, 5, and 9

Real World Problems

These problems apply the skills you've learned in Section 8 to every day situations. As you work through these problems, you'll see that the skills you've learned in this section aren't only important for math tests. They are important skills for the kinds of questions that come up in daily life.

Geometry is logical. When solving a geometry word problem, always look for logical ways to reason your way through the problem. Remember, when solving geometry problems, drawing a figure is often the most helpful tool you can use.

1. How many feet of ribbon will a theatrical company need to tie off a rectangular performance area that is 34 feet long and 20 feet wide?
2. Bridget wants to hang a garland of silk flowers all around the ceiling of a square room. Each side of the room is 9 feet long. The garlands are only available in 15-foot lengths. How many garlands will she need to buy?
3. Pony rides at the local zoo include three trips around a wheel. The wheel around which the ponies walk has a 10-meter spoke (in other words, its radius measures 10 meters). How far does a pony travel during one pony ride?

4. Louise wants to wallpaper her bedroom. It has one window that measures 3 feet by 4 feet and one door that measures 3 feet by 7 feet. The room is 12 feet by 12 feet, and is 10 feet tall. If only the walls are to be covered, rolls of wallpaper are 100 square feet, and no partial rolls can be purchased, what is the minimum number of rolls that she will need?

5. A group of volunteers is searching for a lost camper within a 45-mile radius of the forest ranger's station. What is the total search area in square miles?

6. A builder has 27 cubic feet of concrete to pave a sidewalk whose length is 6 times its width. The concrete must be poured 6 inches deep. How long is the sidewalk?

7. Georgio is making a box. He starts with a 10 × 7-inch rectangle of cardboard. Then, he cuts 2 × 2-inch squares out of each corner. To finish, he folds each side up to make the box, as shown below. What is the box's volume?

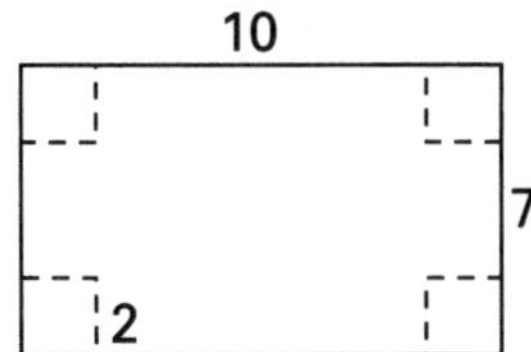

8. Simon needed some cardboard boxes to store some of his things in. A moving company sells boxes in different sizes. Each dimension of the larger box is two inches greater than the same dimension of the smaller box (shown below). How much *more* storage does the large box have than the smaller box?

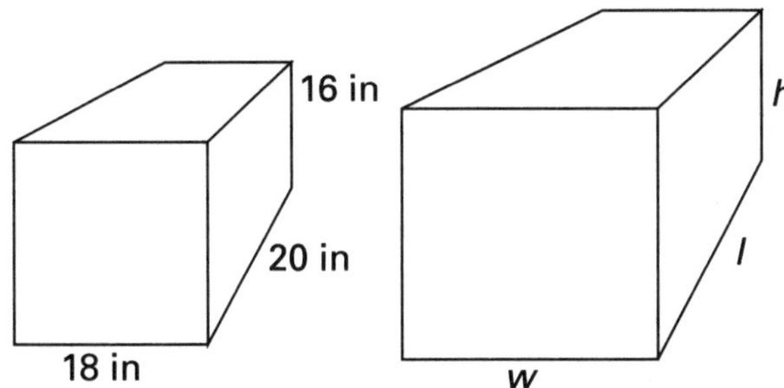

9. Plattville is 80 miles west and 60 miles north of Quincy. How long is a direct route from Plattville to Quincy?

10. During her trip to Yosemite National Park, Macy saw a climber high on a rock wall. At the time, Macy was standing 120 meters from the vertical rock face. Use the diagram below to find out the height of the climber above ground.

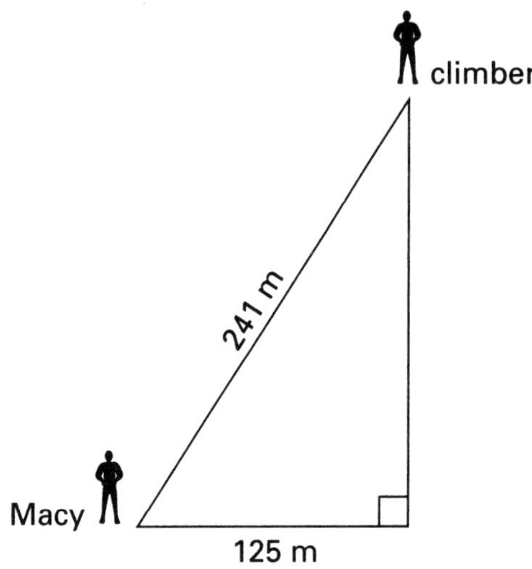

SECTION

8

Answers & Explanations

LESSON 18

1. From the table, you know that this is a ray. So, your answer is $\overrightarrow{QR}$.

2. From the table, you know that this is a line segment. So, your answer is $\overline{MN}$ or $\overline{NM}$.

3. From the table, you know that this is a line. So, your answer is $\overleftrightarrow{GH}$ or $\overleftrightarrow{HG}$.

4. From the table, you know that this is a ray. So, your answer is $\overrightarrow{ZW}$.

5. From the table, you know that this is a line. So, your answer is $\overleftrightarrow{QR}$ or $\overleftrightarrow{RQ}$.

6. From the table, you know that this is a point. So, your answer is "Point P."

7. **a.** $\overleftrightarrow{XY}$, $\overleftrightarrow{YZ}$, $\overleftrightarrow{XY}$, $\overleftrightarrow{XZ}$, $\overleftrightarrow{XZ}$, $\overleftrightarrow{YZ}$

b. From the table you know that there are two ways to name a ray. The endpoint always comes first. $\overrightarrow{YX}$, $\overrightarrow{YZ}$

8. **a.** The rays are $\overrightarrow{JA}$ and $\overrightarrow{JT}$. The vertex is the point that the two rays have in common: J.

b. The rays are $\overrightarrow{YX}$ and $\overrightarrow{YZ}$. The vertex is the point that the two rays have in common: Y.

9. From the table in Think About It, you know that this angle has four different names as follows: ∠KBT, ∠TBK, ∠B, and ∠2.

10. **a.** To measure the angle, begin by placing the cross mark or hole in the middle of the straight edge of the protractor on angle's vertex. Then, line up one of the angle's rays with one of the zero marks on the protractor. Keeping the protractor in place, trace the length of the angle's other ray out to the curved part of the protractor. Read the number on the protractor closest to the ray. The angle measures 45°.

b. To measure the angle, begin by placing the cross mark or hole in the middle of the straight edge of the protractor on angle's vertex. Then, line up one of the angle's rays with one of the zero marks on the protractor. Keeping the protractor in place, trace the length of the angle's other ray out to the curved part of the protractor. Read the number on the protractor closest to the ray. The angle measures 75°.

c. To measure the angle, begin by placing the cross mark or hole in the middle of the straight edge of the protractor on angle's vertex. Then, line up one of the angle's rays with one of the zero marks on the protractor. Keeping the protractor in place, trace the length of the angle's other ray out to the curved part of the protractor. Read the number on the protractor closest to the ray. The angle measures 100°.

d. To measure the angle, begin by placing the cross mark or hole in the middle of the straight edge of the protractor on angle's vertex. Then, line up one of the angle's rays with one of the zero marks on the protractor. Keeping the protractor in place, trace the length of the angle's other ray out to the curved part of the protractor. Read the number on the protractor closest to the ray. The angle measures 125°.

11. **a.** Begin by finding $\overrightarrow{AO}$. Then, find $\overrightarrow{OC}$. Since $\overrightarrow{AO}$ is at 0° and $\overrightarrow{OC}$ is at 70°, you know that ∠AOC measures 70°.

b. Begin by finding $\overrightarrow{AO}$. Then, find $\overrightarrow{OF}$. Since $\overrightarrow{AO}$ is at 0° and $\overrightarrow{OF}$ is at 180°, you know that ∠AOF measures 180°.

c. Begin by finding $\overrightarrow{BO}$. Then, find $\overrightarrow{OD}$. Since $\overrightarrow{BO}$ is at 20° and $\overrightarrow{OD}$ is at 100°, you have to subtract (100 – 20 = 80). So ∠BOD measures 80°.

d. Begin by finding $\overrightarrow{CO}$. Then, find $\overrightarrow{OE}$. Since $\overrightarrow{CO}$ is at 70° and $\overrightarrow{OE}$ is at 135°, you have to subtract (135 – 70 = 65). So ∠COE measures 65°.

12. Begin by looking at the measure of the angle. It's 89°. From the table, you know that an angle that measures less than 90° is called an acute angle. So the angle is acute.

13. Begin by looking at the measure of the angle. It's 180°. From the table, you know that an angle that measures 180° is called a straight angle. So the angle is straight.

14. Begin by looking at the measure of the angle. From the table, you know that an angle with this symbol measures 90° and that a 90° angle is called a right angle. So the angle is right.

15. Begin by looking at the measure of the angle. It's 45°. From the table, you know that an angle that measures less than 90° is called an acute angle. So the angle is acute.

16. Begin by looking at the measure of the angle. It's 150°. From the table, you know that an angle that measures more than 90° and less than 180° is called an obtuse angle. So the angle is obtuse.

17. Begin by looking at the measure of the angle. It's 90°. From the table, you know that an angle that measures 90° is called a right angle. So the angle is right.

18. **a.** Interior angles are those inside the parallel lines. There are two sets of parallel lines in the figure: *a* and *b* and *c* and *d*. For lines *a* and *b*, the interior angles are 2, 3, 6, 7, 10, 11, 14, and 15. For lines *c* and *d*, the interior angles are 5, 6, 7, 8, 9, 10, 11, and 12.

b. Exterior angles are those outside the parallel lines. There are two sets of parallel lines in the figure: *a* and *b* and *c* and *d*. For lines *a* and *b*, the exterior angles are 1, 5, 9, 13, 4, 8, 12, and 16. For lines *c* and *d*, the exterior angles are 1, 2, 3, 4, 13, 14, 15, and 16.

c. Corresponding angles on the same side of the transversal line and either both above or both below the parallel lines. When line *a* is considered the transversal line, $\angle 1$ corresponds to $\angle 9$. When line *c* is considered the transversal line, $\angle 1$ corresponds to $\angle 3$.

d. Alternate interior angles are inside the parallel lines and on opposite sides of the transversal line. When line *a* is considered the transversal line, $\angle 5$ and $\angle 10$ are alternate interior angles. When line *c* is considered the transversal line, $\angle 5$ is an exterior angle.

e. Alternate exterior angles are outside the parallel lines and on opposite sides of the transversal line. When line *c* is considered the transversal line, $\angle 5$ and $\angle 4$ are alternate exterior angles. When line *a* is considered the transversal line, $\angle 5$ is an interior angle.

f. Vertical angles are directly across or opposite from each other. The figure includes the following vertical angle pairs: 1 and 6, 2 and 5, 3 and 8, 4 and 7, 9 and 14, 10 and 13, 11 and 16, and 12 and 15.

g. Supplementary angles add up to 180°. The following angle pairs are supplementary because together they form a straight line: 1 and 2, 3 and 4, 5 and 6, 7 and 8, 9 and 10, 11 and 12, 13 and 14, 15 and 16, 1 and 5, 2 and 6, 3 and 7, 4 and 8, 9 and 13, 10 and 14, 11 and 15, 12 and 16. The following angle pairs are supplementary because they are same side interior angles: 5 and 9, 6 and 10, 7 and 11, 8 and 12, 2 and 3, 6 and 7, 10 and 11, 14 and 15. The following angle pairs are supplementary because they are same side exterior angles: 1 and 13, 2 and 14, 3 and 15, 4 and 16, 1 and 4, 5 and 8, 9 and 12, 13 and 16.

19. You know that $\angle 12$ and $\angle 16$ are supplementary, so together they must add up to 180°. You can find the measure of $\angle 16$ by subtracting the measure of $\angle 12$ from 180. So, $180 - 70 = 110$. Your final answer is 110°.

20. Begin by finding an angle that is related to both angles 1 and 4. You know that $\angle 1$ and $\angle 3$ are congruent because they are corresponding angles. You also know that $\angle 3$ and $\angle 4$ are supplementary because they form a straight line (180°). Then, find the measure of the angle that both angles have in common. You know that $\angle 1$ and $\angle 3$ are congruent. You also know that $\angle 1$ is 120°. So, $\angle 3$ is also 120°. Lastly, find the measure of $\angle 4$ by subtracting: $\angle 4 = 180° - 120° = 60°$. So, your final answer is 60°.

LESSON 19

1. You know that each side of a square is equal. So, if you know that one side of a square is 10 in, then you know that each side of the square is 10 in. To calculate the perimeter, add up the length of all four sides: 10 + 10 + 10 + 10 = 40. So, your final answer is 40 in.
2. Add up the four sides: 3 + 3 + 5 + 5 = 16. So, your final answer is 16 ft.
3. Add up the lengths of the three sides: 3 + 4 + 5 = 12. So, your final answer is 12 mm.
4. You are given the radius, so use $C = 2\pi r$. Plug in the radius and pi and multiply: (2)(3.14)(5) = 31.4. So, your final answer is 31.4 cm.
5. You are given the diameter, so use $C = \pi d$. Plug in the diameter and pi and multiply: (3.14)(10) = 31.4. So, your final answer is 31.4 cm.
6. Add up the lengths of each side: 2 + 7 + 7 + 2 + 10 + 10 = 38. So, the perimeter of the figure is 38 ft.
7. Add up the lengths of each side: 4 + 6 + 2 + 4 + 9 + 5 + 3 = 33. So, the perimeter of the figure is 33 m.
8. Begin by calculating the perimeter of the curved side of the figure. You know that the perimeter of the whole circle is $2\pi r$, or (2)(3.14)(8) = 50.24. Remember, it is a quarter circle, so multiply 50.24 by $\frac{1}{4}$ (or .25) to get 12.56. Both straight edges of the figure are radii, which you are told are 8 cm long each. So the total perimeter of the figure is the curved edge plus the two straight edges: 12.56 + 8 + 8 = 28.56. The final answer is 28.56 cm.
9. Begin by choosing the correct formula. You know this is a triangle, so use $A = \frac{1}{2}bh$. Then, plug in the numbers and solve for the area of the triangle. The base is 7 cm; the height is 6 cm: $A = \frac{1}{2}(6)(7) = 21$. Write your answer in units squared. So, your final answer is 21 cm^2.
10. Begin by choosing the correct formula. From the table, you know this is a parallelogram, so use $A = bh$. Then, plug in the numbers and solve for the area. The base is 12 mm; the height is 10 mm: $A = (10)(12) = 120$. Write your answer in units squared. So, your final answer is 120 mm^2.
11. Begin by choosing the correct formula. You know this is a circle, so use $A = \pi r^2$. Before you can solve for the area of the circle, you have to find the radius. The diameter is 16 inches, so divide by 2 to get the radius: $\frac{16}{2} = 8$. Then, plug in the radius and solve for the area: $A = \pi(8)^2 = 64\pi$. Use $\pi = 3.14$ to get 200.96. Write your final answer in units squared. So your final answer is 200.96 in^2.
12. Begin by choosing the correct formula. From the table you know that this figure is a trapezoid. So, you need to use the following formula: $A = \frac{1}{2}h(b_1 + b_2)$. Then, plug in the measures for the variables in the formula and solve for the missing variable (b_1). It's helpful to first eliminate the parentheses as you learned in the last section.

$$A = \frac{1}{2}h(b_1 + b_2)$$

$$A = \frac{1}{2}hb_1 + \frac{1}{2}hb_2$$

Next, substitute the measures into the formula: $A = 130\text{ cm}^2$; $h = 10$ cm; $b_2 = 16$ cm

$130 = \frac{1}{2}(10)b_1 + \frac{1}{2}(10)(16)$

Now you are ready to solve.

$$130 = \tfrac{1}{2}(10)b_1 + \tfrac{1}{2}(160)$$
$$130 = \tfrac{1}{2}(10)b_1 + 80$$

$$\begin{array}{rl} 130 = & \tfrac{1}{2}(10)b_1 + 80 \\ -80 & \quad -80 \\ \hline 50 = & \tfrac{1}{2}(10)b_1 \\ 50 = & 5b_1 \\ \frac{50}{5} = & \frac{5b_1}{5} \\ 10 = & b_1 \end{array}$$

So, the final answer is 10 cm.

13. First, calculate the area of each circle. To calculate the area of circle A, use $A = \pi r^2$. Plug in the measure of the radius (2) and solve: $A = \pi(2)^2 = 4\pi$. To calculate the area of circle B, again use $A = \pi r^2$. Plug in the measure of the radius (3) and solve: $A = \pi(3)^2 = 9\pi$. Then, subtract the area of circle A from the area of circle B: $9\pi - 4\pi = 5\pi$. Finally, use 3.14 for π: $5\pi = 15.7$. Write the final answer in units squared: 15.7 cm^2.

14. The shaded portion of the figure is the square minus the portion taken up by the circle. So, you first need to calculate the area of the square and the area of the circle. To calculate the area of the square, use $A = bh$. You know that the base and the height have the same value in a square and this will be equal to the diameter of the circle. The radius of the circle is 1, so the diameter of the circle is 2. Thus, the area of the square is (2)(2) = 4. Next, calculate the area of the circle. Use $A = \pi r^2$ where $r = 1$. So, $A = \pi(1)^2 = \pi$. The area of the shaded portion of the figure is equal to the area of the square (4) minus the area of the circle (π, or 3.14). Written in square units, the final answer is 0.86 ft^2.

15. You know that side AB is equal to the diameter of the circle. So begin by calculating the diameter of the circle, using the formula for the area of a circle:
$A = \pi r^2$. Plug in the known measures and solve for r.

$$100\pi = \pi r^2$$
$$100\tfrac{\pi}{\pi} = \tfrac{\pi r^2}{\pi}$$
$$100 = r^2$$

To find the value of r take the square root of each side: $10 = r$. Then, calculate the diameter of the circle. The diameter is twice the radius, or (2)(10) = 20. So, the final answer is 20 cm.

16. Begin by choosing the correct formula. From the table you know that this figure is a rectangular solid and that the formula for its volume is $V = Ah = lwh$. Then, plug in the known measures and solve.

$V = Ah = lwh$

$V = (6)(6)(3)$

$V = 108$

So, the final answer is 108 cm^3.

17. Begin by choosing the correct formula. From the table you know that this figure is a rectangular solid and that the formula for its volume is $V = Ah = lwh$. Then, plug in the known measures and solve.

$V = Ah = lwh$
$V = (12)(4)(9)$
$V = 432$

So, the final answer is 432 in^3.

18. Begin by choosing the correct formula. From the table you know that this figure is a cube and that the formula for the volume of a cube is $V = Ah = s^3$. Plug in the known measures and solve.

$V = Ah = s^3$
$V = 5^3$
$V = 5 \times 5 \times 5$
$V = 125$

So, the final answer is 125 m^3.

19. Begin by choosing the correct formula. From the table you know that this figure is a cylinder and that the formula for its volume is $V = Ah = \pi r^2 h$. Then, plug in the known measures and solve.

$V = \pi r^2 h$
$V = \pi(3)^2(12)$
$V = \pi(9)(12)$
$V = \pi(108)$
$V = 339.12$

So, the final answer is 339.12 m^3.

20. The water fills half the cylinder. So, you could calculate the entire volume of the cylinder and divide by 2. An easier approach might be to directly calculate the volume of the water. To calculate the volume of the water directly, keep in mind that the height of the volume of the water (4 cm) should be used instead of the height of the entire cylinder (8 cm). Begin by choosing the correct formula for the volume of a cylinder: $V = Ah = \pi r^2 h$. Then, plug in the known measures and solve.

$V = \pi r^{2h}$
$V = \pi(5)^2(4)$
$V = \pi(25)(4)$
$V = \pi(100)$
$V = 314$

So, the final answer is 314 cm^3.

LESSON 20

1. Begin by asking yourself: which corresponding parts of the triangles are congruent. You know from the symbols on the triangles that
$\overline{AB} \cong \overline{CD}$
$\overline{BC} \cong \overline{DA}$

In order to prove that the triangles are congruent, you still need one piece of information. Since both triangles share a side ($\overline{AC}$), this side must also be congruent with itself. Side-Side-Side is true, so the two triangles are congruent.

2. Begin by asking yourself: which corresponding parts of the triangles are congruent. You know from the symbols on the triangles that $\overline{MN} \cong \overline{ST}$
$\overline{NL} \cong \overline{TO}$
$\angle N \cong \angle T$
The given information matches the rule for Side-Angle-Side. Since Side-Angle-Side is true, the two triangles are congruent.
3. Begin by asking yourself: which corresponding parts of the triangles are congruent. You know from the symbols on the triangles that $\overline{FG} \cong \overline{IG}$
$\angle F \cong \angle I$
You need to know the measure of $\angle 1$ is congruent to the measure of $\angle 2$. Recall from Lesson 18 that angles directly across from each other (also called vertical angles) are congruent. Using this information, you know that $\angle 1$ is congruent to $\angle 2$. Since Angle-Side-Angle is true, the two triangles are congruent.
4. Begin by asking yourself: which corresponding parts of the triangles are congruent. You know from the symbols on the triangles that
$\angle 1 \cong \angle 4$
$\angle 3 \cong \angle 2$
In order to prove that the triangles are congruent, you still need one piece of information. Since both triangles share a side ($\overline{KM}$), this side must also be congruent with itself. Thus, Angle-Side-Angle is true. The two triangles are congruent.
5. Begin by asking yourself: which corresponding parts of the triangles are congruent. You know from the symbols on the triangles that $\overline{WZ} \cong \overline{YZ}$. Because the two triangles share a side, you know this side is congruent with itself. You also know that $\overline{XZ}$ and $\overline{ZV}$ form a 90° angle, which is supplementary to the angle formed by $\overline{XZ}$ and $\overline{ZW}$. So these two angles must also be congruent. Thus, Side-Angle-Side is true. The two triangles are congruent.
6. Begin by asking yourself which corresponding parts of the triangles are likely to be similar. Since both triangles are right triangles, you know that $\angle C \cong \angle F$. In order for the triangles to be similar $\angle A \cong \angle D$ and $\angle B \cong \angle E$. Since $\angle A$ is 20° and $\angle D$ is 21°, it's not possible for these two triangles to be similar. Therefore, the two triangles are not similar.
7. Begin by asking yourself which corresponding parts of the triangles are likely to be similar. You know from the measurements given that $\angle T \cong \angle L$. From the shapes of the triangles, you know

that $\overline{RT}$ corresponds with $\overline{ML}$ and $\overline{ST}$ corresponds with $\overline{LN}$. (Redrawing one of the figures would make this easier to see.) Then, ask yourself: if one of the rules needed to prove that two triangles are similar is met. To find out if the sides of the triangles are in proportion, you need to set up proportions and see if they are true, as shown below.

$$\frac{8}{15} \stackrel{?}{=} \frac{6}{10}$$

To test whether the proportion is true, you can cross multiply:

$$8 \times 10 \stackrel{?}{=} 15 \times 6$$
$$80 \neq 90$$

Because the two sides of the equation are not equal, the sides are not proportional, and Side-Side-Side is not true. Since none of the rules that prove two triangles are similar are met, the two triangles are not similar.

8. Begin by asking yourself which corresponding parts of the triangles are likely to be similar. Since they share $\angle I$, you know this triangle is congruent. You also know that $\angle H \cong \angle L$ and $\angle J \cong \angle K$. (Remember that corresponding angles formed by a transversal and two parallel lines are congruent.) Because Angle-Angle is true, the two triangles are similar.

9. Begin by asking yourself which corresponding parts of the triangles are likely to be similar. Since $\angle 1$ and $\angle 2$ are vertical angles, you that they are congruent and are both equal to 75°. In order for the triangles to be similar you have to show that either $\angle A \cong \angle D$ or $\angle C \cong \angle E$. Since $\angle 2$ is 75° and $\angle D$ is 20°, you know that $\angle C$ is equal to 85° (180 – 75 – 20 = 85). This proves that both $\angle A \cong \angle D$ and $\angle C \cong \angle E$. Therefore, the triangles are similar according to Angle-Angle.

10. You know from the problem that the triangles are similar. So you can set up a proportion to solve for the height of the flagpole.

$$\frac{\text{Person's height}}{\text{Flagpole's height}} = \frac{\text{Length of the person's shadow}}{\text{Length of the flagpole's shadow}}$$

Since you are looking for the height of the flagpole, let x = the height of the flagpole. Then, plug in the measures and solve for x.

$$\frac{6}{x} = \frac{4}{20}$$

Cross multiply to solve for x.

$$6 \times 20 = 4x$$
$$120 = 4x$$
$$\frac{120}{4} = \frac{4x}{4}$$
$$30 = x$$

So, the flagpole is 30 feet tall.

LESSON 21

1. $\overline{XY}$
2. $\overline{HP}$ and $\overline{PT}$
3. $\overline{YZ}$ and $\overline{XZ}$
4. $\overline{HT}$
5. Begin by finding the legs and the hypotenuse in the right triangle. The legs are $\overline{DF}$ and $\overline{FE}$. The hypotenuse is $\overline{DE}$. Then, match up the variables *a*, *b*, and *c* with the parts of the right triangle. For example, you might

   Let a = the length of $\overline{DF}$.
   Let b = the length of $\overline{FE}$.
   Let c = the length of $\overline{DE}$.

   Next, plug the known lengths into the Pythagorean Theorem and solve.

   $a^2 + b^2 = c^2$
   $8^2 + 15^2 = c^2$
   $64 + 225 = c^2$
   $289 = c^2$
   $\sqrt{289} = \sqrt{c^2}$
   $17 = c$

   So, the length of the hypotenuse is 17.
6. Begin by finding the legs and the hypotenuse in the right triangle. The legs are $\overline{JL}$ and $\overline{LK}$. The hypotenuse is $\overline{JK}$. Then, match up the variables *a*, *b*, and *c* with the parts of the right triangle. For example, you might

   Let a = the length of $\overline{JL}$.
   Let b = the length of $\overline{LK}$.
   Let c = the length of $\overline{JK}$.

   Next, plug the known lengths into the Pythagorean Theorem and solve.

   $a^2 + b^2 = c^2$
   $a^2 + 12^2 = 15^2$
   $a^2 + 144 = 225$

   $a^2 + 144 = 225$
   $\quad - 144 \quad - 144$
   $a^2 = 81$
   $\sqrt{a^2} = \sqrt{81}$
   $a = 9$

   So, the length of the missing leg is 9.

7. Begin by finding the legs and the hypotenuse in the right triangle. The legs are $\overline{RT}$ and $\overline{TS}$. The hypotenuse is $\overline{RS}$. Then, match up the variables *a*, *b*, and *c* with the parts of the right triangle. For example, you might

Let a = the length of $\overline{RT}$.
Let b = the length of $\overline{TS}$.
Let c = the length of $\overline{RS}$.

Next, plug the known lengths into the Pythagorean Theorem and solve.

$$a^2 + b^2 = c^2$$
$$6^2 + b^2 = 10^2$$
$$36 + b^2 = 100$$
$$\begin{array}{r} 36 + b^2 = 100 \\ -36 \qquad -36 \\ \hline \end{array}$$
$$b^2 = 64$$
$$\sqrt{b^2} = \sqrt{64}$$
$$b = 8$$

So, the length of the missing leg is 8.

8. If the triangle with sides, 5, 6, and 9 is a right triangle, its sides will follow the pattern of the Pythagorean Theorem, where $a^2 + b^2 = c^2$, where c is the hyponteneuse. You just need to plug the given variables into the equation and solve:

$$a^2 + b^2 = 9^2$$
$$5^2 + 6^2 \stackrel{?}{=} 9^2$$
$$25 + 36 \neq 81$$
$$\sqrt{61} = \sqrt{c^2}$$

Because the legs do not follow the Pythagorean Theorem, the triangle is not a right triangle.

9. If the triangle with sides 6, 8, and 10 is a right triangle, its sides will follow the pattern of the Pythagorean Theorem where $a^2 + b^2 = c^2$, where c is the hypoteneuse. So, you plug the variables into the equation and see if the math works out.

$$a^2 + b^2 = c^2$$
$$6^2 + 8^2 \stackrel{?}{=} 10^2$$
$$36 + 64 = 100$$

Because the legs follow the Pythagorean Theorem, you know that it's a right triangle.

10. If the triangle with sides 3, 5, and 9 is a right triangle, its sides will follow the pattern of the Pythagorean Theorem, where $a^2 + b^2 = c^2$, where c is the hypoteneuse. So you plug the variables into the equation and solve:

$$a^2 + b^2 = c^2$$
$$3^2 + 5^2 \stackrel{?}{=} 9^2$$
$$9 + 25 \neq 81$$

Because the legs do not follow the Pythagorean Theorem, you know that it's not a right triangle.

REAL WORLD PROBLEMS

1. Begin by determining what you need to find out. The performance area will form a rectangle. The ribbon will outline the rectangle, so you need to find the perimeter of the performance area. To find the perimeter of the rectangle, add up the lengths of the four sides: 34 + 20 + 34 + 20 = 108. So, 108 feet of ribbon will be needed.
2. Begin by calculating the perimeter of the room. Since the room is a square, you can calculate its perimeter using the formula $4s$, where s = the length of one side of the room, or 9 feet. So, the perimeter is $4 \times 9 = 36$ feet long. Bridget will need three garlands to cover the perimeter of the room because 2 garlands ($2 \times 15 = 30$) will be too short. Your final answer is three garlands.
3. Begin by visualizing the situation. The ponies walk in a circle. The radius of the circle is equal to the length of the wheel spoke, or 10 meters. So, first you need to calculate the perimeter, or circumference, of the circle that the pony walks in one turn of the wheel. The formula for the circumference of a circle is $C = 2\pi r$. Plug in the numbers and multiply: $C = (2)(3.14)(10) = 62.8$. Now you are ready to calculate the distance a pony travels in one ride: $(3)(62.8) = 188.4$. So, one pony ride is 188.4 meters long.
4. Begin by calculating the area of the walls that needs to be covered. First, calculate the area of the walls. The room has four walls, each 12 feet long and 10 feet tall, so the area of each wall is 120 square feet ($10 \times 12 = 120$). There are four walls, so the area of all four walls is 480 square feet ($4 \times 120 = 480$). Only the walls will be covered with wallpaper, so you need to subtract the area that the window and the door take up in the room. The window takes up 12 square feet ($3 \times 4 = 12$). The door takes up 21 square feet ($3 \times 7 = 21$). So, the total area that Louise needs to cover with wallpaper is $480 - 12 - 21 = 447$ square feet. Since each roll of wallpaper covers 100 square feet, she will need to buy five rolls of wallpaper.
5. Think about which area formula will work in this problem. The problem tells you the radius, which gives you the clue that you are looking for the area of a circle. The formula for the area of a circle is $A = r^2$. So, plug in the numbers and solve:

$$\begin{aligned} A &= \pi r^2 \\ &= (3.14)(45)^2 \\ &= (3.14)(2{,}025) \\ &= 6{,}358.5 \end{aligned}$$

So, the total search area is 6,358.5 square miles.

6. The sidewalk is a rectangular solid and you are asked to determine its length. You know that the length is six times longer than the width. Let x = the width of the sidewalk. Then, use the formula for the volume of a rectangular solid to solve for the sidewalk's length. The formula for the volume of a rectangular solid is $V = Ah = lwh$. So,

$$V = Ah = lwh$$
$$27 = (6x)(x)(0.5)$$
$$27 = (6x^2)(0.5)$$
$$\frac{27}{3} = \frac{3x^2}{3}$$
$$9 = x^2$$
$$3 = x$$

This tells you that the width of the sidewalk is 3 feet. The length of the sidewalk is six times longer, or 18 feet long ($3 \times 6 = 18$). So, your final answer is 18 feet long.

7. You are looking for the volume of a rectangular solid. The formula for the volume of a rectangular solid is $V = Ah = lwh$. The length is 10 minus 2 inches on each side, or $10 - 4 = 6$ inches long. The width is 7 inches minus 2 inches on each side, or $7 - 2 - 2 = 3$. The height is 2 inches.

$$V = Ah = lwh$$
$$= (6)(3)(2)$$
$$= (18)(2)$$
$$= 36$$

So, the box's volume is 36 cubic inches.

8. Begin by calculating the volume of each box shown. Both boxes are rectangular solids. The formula for the volume of a rectangular solid is $V = Ah = lwh$. To calculate the volume of the smaller box, plug in the dimensions shown in the figure.

$$V = Ah = lwh$$
$$= (20)(18)(16)$$
$$= (360)(16)$$
$$= 5{,}760$$

So, the box's volume is 5,760 cubic inches.

To calculate the volume of the larger box, add 2 inches to each dimension shown in the figure.

$$V = Ah = lwh$$
$$= (22)(20)(18)$$
$$= (440)(18)$$
$$= 7{,}920$$

So, the box's volume is 7,920 cubic inches.

The difference in the two volumes is 2,160 cubic inches (7,920 – 5,760 = 2,160). So your final answer is 2,160 cubic inches greater storage volume.

9. The distances given form a right triangle, so you can use the Pythagorean Theorem to solve for the missing distance. The missing distance will form the hypotenuse of the triangle because it comes directly across from the right angle. So, plug the variables into the equation and solve for c:

$$a^2 + b^2 = c^2$$
$$60^2 + 80^2 = c^2$$
$$3{,}600 + 6{,}400 = c^2$$
$$10{,}000 = c^2$$
$$\sqrt{10{,}000} = \sqrt{c^2}$$
$$100 = c$$

So, a direct route from Plattville to Quincy is 100 miles.

10. The distances form a right triangle, so you can use the Pythagorean Theorem ($a^2 + b^2 = c^2$). The distance from Macy to the rock wall is 120 m, so $a = 120$. The hypotenuse of the triangle in the diagram is 241 m, so $c = 241$. Using these values yields the equation $14{,}400 + b^2 = 58{,}081$.

Now, solve the equation for b:

$$b^2 = 58{,}081 - 14{,}400$$
$$b^2 = 43{,}681$$

Then, you take the square root of both sides: $\sqrt{b^2} = \sqrt{43{,}681}$

So, your final answer is $b = 209$. The climber was 209 meters off the ground.

Check Your Understanding

POSTTEST

Now that you've worked through all the lessons in this book, you're ready to show off your math skills by taking this Posttest. The Posttest is 25 multiple choice questions that cover the basic concepts and skills covered in this book. Even though the Posttest looks very similar to the Pretest, the questions are all different.

You can use the answer sheet on the next page to record your answers to the Posttest. Or, you can simply circle your answers as you work through the Posttest. If this book doesn't belong to you, write the numbers 1–25 on a sheet of your own paper and record your answers there.

Take your time to work through the problems. Don't feel rushed. Once you've completed the Posttest, check your answers against the answer key at the end of the test. Then, compare your score on the Pretest with your score on this Posttest.

How did you do? If you scored better on the Posttest than on the Pretest—Congratulations! Your hard work shows! If there are areas where you can still improve, review those lessons again. Each question is keyed to the lessons you need to review to get the answer correct. Math skills are lifelong—you don't have to learn them all today. Keep working on the areas that are difficult for you, and you'll see continued improvement—at school and in other parts of your life.

POSTTEST ANSWER SHEET

1. ⓐ ⓑ ⓒ ⓓ
2. ⓐ ⓑ ⓒ ⓓ
3. ⓐ ⓑ ⓒ ⓓ
4. ⓐ ⓑ ⓒ ⓓ
5. ⓐ ⓑ ⓒ ⓓ
6. ⓐ ⓑ ⓒ ⓓ
7. ⓐ ⓑ ⓒ ⓓ
8. ⓐ ⓑ ⓒ ⓓ
9. ⓐ ⓑ ⓒ ⓓ
10. ⓐ ⓑ ⓒ ⓓ
11. ⓐ ⓑ ⓒ ⓓ
12. ⓐ ⓑ ⓒ ⓓ
13. ⓐ ⓑ ⓒ ⓓ
14. ⓐ ⓑ ⓒ ⓓ
15. ⓐ ⓑ ⓒ ⓓ
16. ⓐ ⓑ ⓒ ⓓ
17. ⓐ ⓑ ⓒ ⓓ
18. ⓐ ⓑ ⓒ ⓓ
19. ⓐ ⓑ ⓒ ⓓ
20. ⓐ ⓑ ⓒ ⓓ
21. ⓐ ⓑ ⓒ ⓓ
22. ⓐ ⓑ ⓒ ⓓ
23. ⓐ ⓑ ⓒ ⓓ
24. ⓐ ⓑ ⓒ ⓓ
25. ⓐ ⓑ ⓒ ⓓ

1. Pedro wanted to find the volume of his rabbit's rectangular hutch. He measured the dimensions of the hutch and got a volume of 31.32 cubic feet. Which of the following could be the dimensions of the rabbit's hutch?
   a. 2.7 ft × 2.9 ft × 4 ft
   b. 2.4 ft × 2.8 ft × 3.9 ft
   c. 3 ft × 4 ft × 2.5 ft
   d. 2 ft × 2 ft × 8 ft

2. What fraction of the figure is shaded in?

   a. $\frac{5}{16}$
   b. $\frac{1}{2}$
   c. $\frac{3}{8}$
   d. $\frac{3}{4}$

3. Convert $1\frac{5}{7}$ to an improper fraction.
   a. $\frac{12}{7}$
   b. $\frac{6}{7}$
   c. $\frac{2}{14}$
   d. $\frac{5}{14}$

4. A recipe calls for $\frac{1}{2}$ teaspoon of coriander and serves four people. If Heather wants to serve the recipe to 24 people, how much coriander will she need?
   a. 3 teaspoons
   b. $3\frac{1}{2}$ teaspoons
   c. 4 teaspoons
   d. 6 teaspoons

5. 6.7 – 2.135 =

a. 4.656
b. 0.4656
c. 0.4565
d. 4.565

6. The amount $480.85 is to be divided equally among four people. To the nearest cent, how many dollars will each person get?

a. $125.25
b. $122.23
c. $120.21
d. $120.20

7. Which of the following is another way to write 27.5%?

a. $\frac{0.275}{100}$
b. $\frac{23}{80}$
c. $\frac{11}{40}$
d. $\frac{7}{20}$

8. Which of the following is 400% of 30?

a. 1.2
b. 12
c. 120
d. 1,200

The pie chart below shows Liam's monthly expenses. Use this information to answer Question 9.

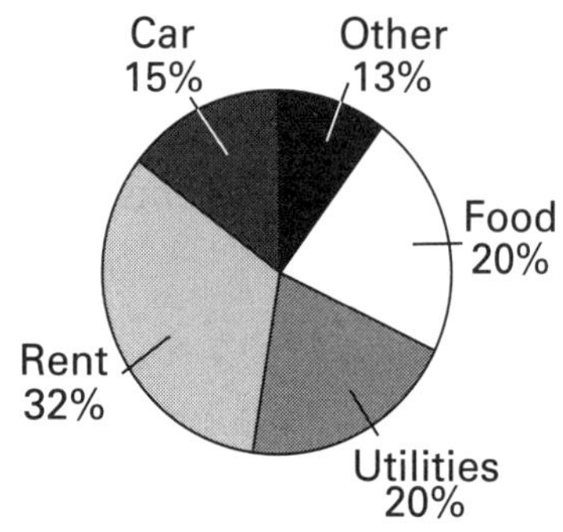

9. If Liam's total expenses are $2,000 per month, how much does he spend on his car each month?

a. $220
b. $300
c. $320
d. $360

10. Which of the following means the same as $3x - 8 = 25$?

a. three less than eight times a number is 25
b. eight more than three times a number is 25
c. three more than eight times a number is 25
d. eight less than three times a number is 25

11. A number is decreased by 5, then decreased by 13, resulting in 41. What is the original number?

a. 23
b. 59
c. 62
d. 70

12. $8 \times 3 + 20 \div 4 =$

a. 11
b. 13
c. 19
d. 29

13. Which of the following is equal to 41^2?

a. 41×41
b. 1,600
c. $\sqrt{41}$
d. $\sqrt{41} \times \sqrt{41}$

14. What is the measure of ∠ROQ?

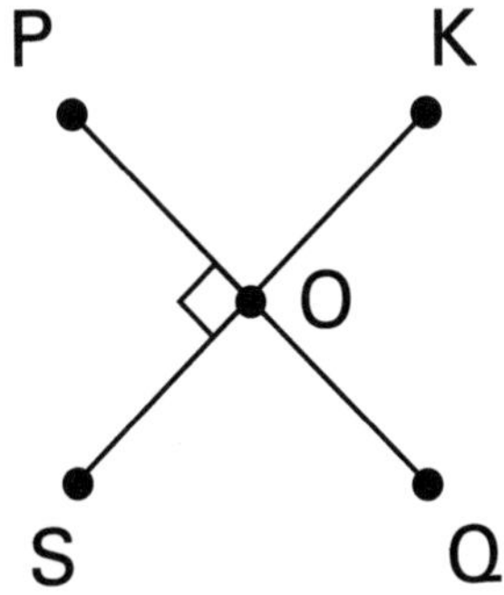

a. 30°
b. 45°
c. 90°
d. 180°

15. A 10-foot ladder is leaning against a building as shown in the figure below. How many feet off the ground is the top of the ladder?

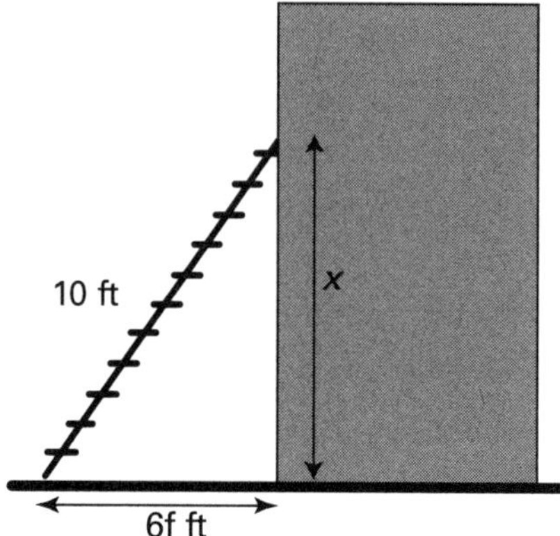

a. 8
b. 10
c. 12
d. $\sqrt{8}$

16. The height of the Statue of Liberty from foundation to torch is 305 feet. A mini-golf park has a 1:60 scale model of the statue. Approximately, how tall is the scale model?

a. 3 feet
b. 5 feet
c. 10 feet
d. 12 feet

17. The table below shows the how many male and female students participate in each activity at Martin Luther King, Jr., Middle School. What is the ratio of males to females on the Debate Team?

ACTIVITY	MALE	FEMALE
Drama	11	13
Journalism	12	10
Science Club	9	11
Debate Team	12	15

a. $\frac{11}{13}$
b. $\frac{12}{10}$
c. $\frac{15}{12}$
d. $\frac{12}{15}$

18. There are 36 blue marbles, 16 white marbles and 28 green marbles in a bag. If one marble is drawn from the bag, what is the probability it is a white marble?

a. $\frac{1}{5}$

b. $\frac{1}{16}$

c. $\frac{1}{32}$

d. $\frac{1}{64}$

19. Malaika has ten grades of equal weight in Math. They are 83, 85, 75, 90, 88, 86, 91, 93, 82, and 85. What is Malaika's average in Math? (Round your answer to the nearest whole number.)

a. 82

b. 84

c. 86

d. 88

20. $9\frac{5}{6} - 7\frac{1}{8} =$

a. $\frac{29}{24}$

b. $2\frac{17}{24}$

c. $2\frac{5}{8}$

d. $\frac{13}{12}$

21. Triangles RST and MNO are similar. What is the length of $\overline{MO}$?

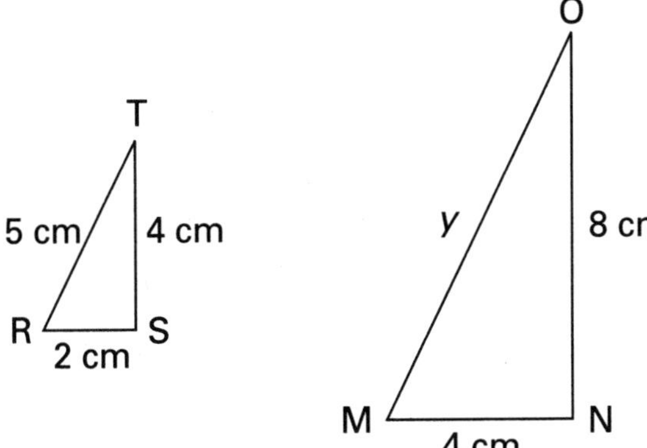

a. 5 cm

b. 10 cm

c. 20 cm

d. 32 cm

22. Emmanuel used data from a local pet store to compare the amounts of dog food different breeds eat in one month. He recorded his data in the following graph.

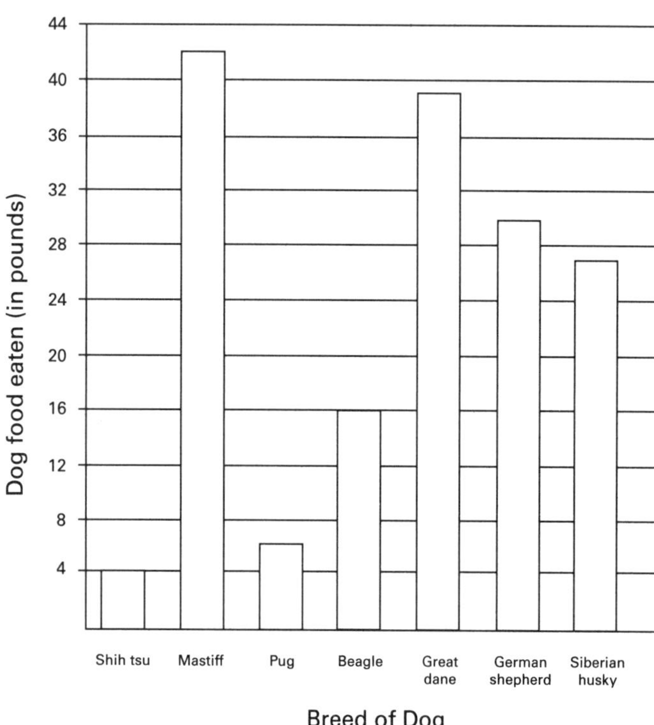

How much more food did the Great Dane eat than the beagle?

a. 10 pounds
b. 23 pounds
c. 39 pounds
d. 55 pounds

23. There are 21 people in Mao's History class. If there is a ratio of boys to girls of 3:4, how many boys are in Mao's History class?

a. 3
b. 6
c. 9
d. 12

24. Last season, the Bulldogs won 32 games. They lost only 4 games. There were no tied games. What is the ratio of games won to games played?

a. 15%
b. 4:32
c. 1:9
d. 8:9

25. Which of the measures represents an obtuse angle?

a. 60°

b. 85°

c. 90°

d. 105°

POST-TEST ANSWERS

1. a. You know that you need to multiply three dimensions to get the volume of something, so you multiply the numbers in the answer choices together in order to see which number matches Pedro's answer. You can begin by multiplying each set of dimensions given in the answer choices: Choice **a** is: 2.7 ft × 2.9 ft × 4 ft, which equals 31.32 ft^3. Since answer choice **a** matches the volume given in the question, you don't need to multiply the other answer choices. If you missed this question, you might want to review Lesson 19.

2. c. Because 6 out of 16 total parts are shaded, you know that $\frac{6}{16}$ of the figure is shaded. This fraction can be reduced to lowest terms by dividing the numerator and the denominator by 2. So your final answer is $\frac{3}{8}$. If you missed this question, you might want to review Lesson 3.

3. a. Multiply the whole number (1) by the denominator (7). Then, you add the numerator (5), and keep the same denominator.

$1\frac{5}{7}$ (1 times 7, plus 5)

So, your final answer is $\frac{12}{7}$. If you missed this question, you might want to review Lesson 6.

4. a. Heather wants to serve six times more people than the original recipe is designed to serve (24 ÷ 4 = 6). So, she will need to add six times more coriander than the recipe lists: $6 \times \frac{1}{2} = 3$. So, the final answer is 3 teaspoons. Multiplying fractions is covered in Lesson 5. You could also solve this question using a proportion. Proportions are covered in Lesson 12.

5. d. To solve this problem correctly, you must be sure that the decimal points are lined up correctly. Then, subtract:

$$\begin{array}{r} 6.7 \\ -\,2.135 \\ \hline 4.565 \end{array}$$

So, your final answer is 4.565. If you missed this question, you might want to review Lesson 7.

6. c. Divide the total amount ($480.85) by four people. To do this, you need to first set up the division problem. Begin by writing the decimal point in the answer portion of the division problem. It should go just above the decimal point in the problem. Now you are ready to divide:

$$4\overline{)480.85}$$

The answer is 120.2125. The problem asks you to report your final answer rounded to the nearest cent, so you need to round this decimal to the nearest hundredths place. Because the number in the thousandths place is less than 5, the answer is rounded to 120.21. So, your final answer is \$120.21. If you missed this problem, you might want to review Lessons 7 and 8.

7. c. Begin by writing the percent as a fraction over the denominator 100: $\frac{27.5}{100}$. This is not an answer choice, so you need to reduce the fraction. You can simplify the problem by multiplying by $\frac{10}{10}$ to get $\frac{275}{1,000}$. Then reduce as follows: $\frac{275}{1,000} = \frac{55}{200} = \frac{11}{40}$. So, the final answer is $\frac{11}{40}$. If you missed this problem, you might want to review Lesson 9.

8. c. Begin by figuring out what you know from the problem and what you're looking for.
- ▶ You have the percent: 400%.
- ▶ You have the whole: 30.
- ▶ You are looking for the part.

Then, use the following equation to solve the problem: Whole × Percent = Part. Plug in the parts of the equation that you know: Part = 30 × 4.00 = 120. The answer is 120. If you missed this question, you might want to review Lesson 10.

9. b. According to the graph, 15% of the total monthly expenses goes toward Liam's car, so you need to find 15% of \$2,000. Fifteen percent is equal to 0.15 and *of* means to multiply: 0.15 × 2,000 = 300. So, Liam spends \$300 on his car each month. If you missed this question, you might want to review Lesson 14.

10. d. Break the math symbols down as follows:

$$3x - 8 = 25$$

eight less than

three times a number

equals 25

So, the answer is eight less than three times a number is 25. If you missed this question, you might want to review Lessons 16 and 17.

11. b. Begin by letting x = the number you are looking for. Then, set up an algebraic equation and solve for x. From the problem, you know that first 5 is subtracted from the number, then 13 is subtracted from the number, so your equation should be:

$$x - 5 - 13 = 41$$

First, you combine like terms.

$$x - 18 = 41$$

Then, you add 18 to each side of the equation and solve for x:

$$\begin{array}{r} x - 18 = 41 \\ +18 + 18 \\ \hline x = 59 \end{array}$$

So, the final answer is 59. If you missed this question, you might want to review Lesson 17.

12. d. To solve this problem, you simply follow the order of operations. First, multiply and divide numbers in order from left to right.

$$8 \times 3 + 20 \div 4 =$$
$$24 + 5 =$$

Then, add and subtract numbers in order from left to right.

$24 + 5 = 29$

So the final answer is 29. If you missed this question, you might want to review Lesson 1.

13. a. When a number is followed by a raised 2, you square it, which is the same as multiplying it by itself. Thus, $41^2 = 41 \times 41$. If you missed this question, you might want to review Lesson 2.

14. c. You know that $\angle POS$ measures 90° because it is marked as a right angle. $\angle ROQ$ is opposite, or vertical to, $\angle POS$, so it also measures 90°. If you missed this question, you might want to review Lesson 18.

15. a. The ladder forms a right triangle with the ground and the side of the building, so you can use the Pythagorean Theorem to solve for the missing distance. Plug the variables into the equation and solve for a:

$$a^2 + b^2 = c^2$$
$$a^2 + 6^2 = 10^2$$
$$a^2 + 36 = 100$$

$$\begin{array}{l} a^2 + 36 = 100 \\ \quad - 36 \;\; - 36 \\ \hline a^2 \quad = 64 \\ \sqrt{a^2} = \sqrt{64} \\ a \quad = 8 \end{array}$$

So, the top of the ladder is about 8 feet above the ground. If you missed this question, you might want to review Lesson 21.

16. b. Begin by setting up a proportion:

$$\frac{1}{60} = \frac{\text{height of the mini-golf model of the statue}}{\text{actual height of the Statue of Liberty}} = \frac{x}{305}$$

Then, cross multiply and solve for x:

$$60x = 305$$

$$\frac{60x}{60} = \frac{305}{60}$$

$$x = 5.08$$

The answer choices given are all rounded to the nearest foot and the problem uses the word *approximately*, so the answer is 5 feet. If you missed this question, you might want to review Lesson 12.

17. d. The question asks you to find the ratio of males to females on the Debate Team. Go to the row labeled Debate Team in the table. Then, set up a ratio comparing the number of males to females: $\frac{12}{15}$. If you missed this question, you might want to review Lessons 12 and 15.

18. a. There are 16 white beads (favorable outcomes) and 80 total beads (possible outcomes). Plug these numbers into the equation for probability:

$$\text{P (event)} = \frac{\text{number of favorable outcomes}}{\text{number of total outcomes}} = \frac{16}{80}$$

Therefore, the probability of drawing a white bead is $\frac{16}{80}$, which reduces to your final answer, $\frac{1}{5}$. If you missed this question, you might want to review Lesson 13.

19. c. To find the average, you need to add Malaika's grades together and divide by 10:

$$\frac{83 + 85 + 75 + 90 + 88 + 86 + 91 + 93 + 82 + 85}{10} = 85.8.$$

When dividing by a multiple of 10, you just have to move the decimal point.

Thus, Malaika's average in Math is 85.8. The question tells you to round your answer to the nearest whole number, so you need to round your answer to 86. If you missed this question, you might want to review Lesson 11.

20. b. First, convert the fractions so that they have a lowest common denominator of 24: $\frac{5}{6}$ becomes $\frac{20}{24}$, and $\frac{1}{8}$ becomes $\frac{3}{24}$. Then, subtract the fractions $\frac{20}{24} - \frac{3}{24} = \frac{17}{24}$. Next, subtract the whole numbers: $9 - 7 = 2$. Now combine the fraction and the whole number: $2\frac{17}{24}$. This is your final answer. If you missed this question, you might want to review Lessons 3, 4, and 6.

21. b. The sides of similar triangles are proportional. So, you can set up a proportion to solve for the missing length.

$\frac{2}{4} = \frac{5}{y}$

Cross multiply to solve for y.

$4 \times \frac{2y}{4} = 5 \times 4$

$2y = 20$

$\frac{2y}{2} = \frac{20}{2}$

$y = 10$

So, $\overline{MO}$ is 10 cm. If you missed this question, you might want to review Lesson 20.

22. b. Subtract the amount the beagle ate from the amount the Great Dane ate: 39 – 16 = 23. The Great Dane ate 23 more pounds of food than did the beagle. If you missed this question, you might want to review Lesson 14.

23. c. First, set up the proportion:

$\frac{3}{7} = \frac{?\text{ boys}}{21}$

Then, solve for the missing number by cross multiplying and solving for the missing number. (Let x = the missing number.)

$3 \times 21 = 7x$

$63 = 7x$

$\frac{63}{7} = \frac{7x}{7}$

$9 = x$

So, there are nine boys in Mao's class. If you missed this question, you might want to review Lesson 12.

24. d. The question asks for the ratio of games won to games played. First, you need to calculate the total number of games played. Since the Bulldogs won 32 games, lost 4 games, and tied no games, they must have played a total of 36 games. The ratio of games won to games played is

$\frac{32\text{ games won}}{36\text{ total games}}$, or $\frac{32}{36}$.

Reduce $\frac{32}{36}$ to your final answer, $\frac{8}{9}$. If you missed this question, you might want to review Lesson 12.

25. d. Obtuse angles are greater than 90°. Only one answer choice (**d**) is greater than 90°. If you missed this question, you might want to review Lesson 18.